The Broadcaster's Guide to RDS

Scott Wright

Focal Press
Boston   Oxford   Johannesburg   Melbourne   New Delhi   Singapore

Library of Congress Cataloging-in-Publication Data
Wright, Scott.
    The broadcaster's guide to RDS / by Scott Wright.
       p.   cm.
    Includes bibliographical references (p.   ) and index.
    ISBN 0-240-80278-0 (pbk. : alk. paper)
    1. Radio frequency modulation—Standards. 2. Teletext systems.
I. Title.
TK6553.W69   1997
384.54'4—dc21   97-5788
                                                        CIP

British Library Cataloguing-in-Publication Data
A catalogue record for this book is available from the British Library.

The publisher offers special discounts on bulk orders of this book.
For information, please contact:
Manager of Special Sales
Butterworth-Heinemann
225 Wildwood Avenue
Woburn, MA 01801-2041
Tel: 617-928-2500
Fax: 617-928-2620

For information on all Focal Press publications available, contact our World Wide Web home page at: http://www.bh.com/focalpress

10 9 8 7 6 5 4 3 2 1

Transferred to Digital Printing 2006

For my family, friends, and colleagues
You are what is important to me
And what makes my life meaningful
Thanks for your guidance and understanding

You can surrender without a prayer
But never really pray
Without surrender

You can fight
Without ever winning
But never ever win
Without a fight

Neil Peart

Table of Contents

Introduction

On January 3, 1993, the National Radio Systems Committee (NRSC) approved the *Radio Broadcast Data System (RBDS) Standard* for use by FM broadcasters. This subcarrier data system is unlike any other subcarrier service available today, either audio or data, in that it provides a data system for the broadcaster to communicate with his listening public. The information contained is relevant only to that station and is used to provide a variety of tuning and information services. RDS changes the way the listener interacts with his radio unlike any other technology.

The *RBDS Standard* is very closely crafted after the proven and widely-used *Radio Data System (RDS) Standard* that is used in Europe. People often use RDS and RBDS synonymously. Smart Radio is a term coined for use in the United States as well. Almost every broadcaster in Europe and in many other countries around the world transmit RDS data. The *RDS Standard* was approved by the European Broadcasting Union (EBU) in 1989 and has proven its usefulness to both the broadcaster and the listener. In fact, nearly 60% of all receivers sold in Europe are equipped with RDS capability.

Since the approval of the U.S. standard, over 600 stations have begun broadcasting RDS. There is a wide variety of encoding equipment available today, with prices starting as low as $1000. Receiver manufacturers have introduced over thirty different models of RDS receivers. These include car and home receivers, home tuners, home theater, portable, and even receivers for the personal computer. Even the FCC is aware of the power of RBDS to reach the mass consumer and strongly encourages the use of RBDS by FM broadcasters with the new Emergency Alerting System (EAS). But even with all the positive support for RBDS, the individual who can most benefit from its use, the FM broadcaster, has been reluctant to implement and utilize this service. The reason for this reluctance is mainly confusion and a lack of understanding of what RBDS can do for your station. This book will present practical methods of implementing RDS and reveal how your listeners will now interact with your broadcast in new and exciting ways.

Since the RDS and RBDS standards were approved, a number of requests have been made for enhancements and improvements of the system. Many of these requests dealt with the use of RDS as a datacasting technology. With a great deal of cooperation between the NRSC RBDS subcommittee and the European RDS forum, both standards have been upgraded and revised with an eye towards global harmonization. While subtle differences still remain, the two standards are as identical as possible. In fact, the *RBDS Standard* includes the entire *RDS Standard*, with the addition of features necessary for the North American market. At the time of this writing, both standards are in the final drafting and approval process, and final approval for both standards is expected sometime in 1997. These changes are covered in detail in this book.

The purpose of this book is to provide the information required to understand RBDS on a variety of levels so that everyone involved in radio can make the most of the system. This book is for the station owner, program director, salesman, radio personality, and engineer alike. I have organized the chapters such that you may start at the beginning and read through until you have reached the desired level of expertise you require in your job to fully utilize this new technology. By the way, everyone is encouraged to read past this introduction!

1

What is RDS, or RBDS, or Smart Radio?

You may have heard any of the above terms; they all basically refer to the same thing. The definitions are as follows:

- RDS—Radio Data System. This is the standard utilized everywhere in the world except North America. All aspects of the RDS standard are encompassed within the RBDS standard.
- RBDS—Radio Broadcast Data System. This technical standard is exactly like the RDS standard, except that additional features have been added that are tailored to the needs of the American broadcaster. RBDS is the name of the standard only. All references should otherwise refer to the technology as RDS.
- Smart Radio. A termed coined by the receiver manufacturer Denon, so that we wouldn't have to remember RDS or RBDS. This is a catchy and descriptive way to market this technology to the consumer.

For the purposes of this book I will use the term RDS. Now, then, RDS is a 57 kHz data subcarrier that transmits advanced radio tuning and program information from the broadcaster to the listener. This subcarrier is injected into the composite signal by use of an RDS encoder. RDS is used only on the FM band and does not require that the station broadcast in stereo. The data rate is approximately 1200 baud (bits per second) and typically requires only 2.0% modulation. The data is transmitted in a continuous cyclic fashion, resulting in a very robust datastream. The data is organized into functional groups so that only pertinent data is transmitted. Information that

1

must always be transmitted appears in every data group for easy decoding by the receiver.

The broadcaster typically enters information into the encoder with the use of a computer and a serial RS-232 link. Most manufacturers provide interface software that runs on an IBM or IBM-compatible computer. If you do not own a computer, you may purchase an encoder with the necessary information already stored into non-volatile memory by the manufacturer. Most encoders offer storage of multiple sets of information typically referred to as data records. Some encoders offer the ability to change the transmitted data record by means of a push-button interface. In all cases, if you need either to enter information that is not already stored or to change previously stored information, then a computer is necessary. If you are constantly entering in new information, then a dedicated personal computer is required. The true success of any information service is that the more dynamic and relevant the information is, the more interesting it is for the listener. It is to your long-term benefit to go "the extra mile" and provide dynamic RDS data to your best asset—your listener.

For the consumer to use the RDS information being broadcast, he or she must purchase a specially-equipped receiver. There are no add-ons or black boxes currently being marketed. Also, for the automated tuning features to be used, RDS must be integrated into the receiver. The RDS receiver uses an RDS decoder Integrated Circuit (IC) that demodulates the subcarrier data into a continuous stream of Logic 1's and 0's, then passes that stream to a microprocessor. Microprocessors are used in all electronically-tuned receivers today. The microprocessor then decodes the digital information back into its functional group for use by the receiver-tuner, or displays it to the consumer via an alphanumeric display. RDS receivers typically have additional push buttons on the front panel that allow the consumer to select and activate certain features. Just like any other technology, the added cost is proportional to the number of added features. A general rule of thumb is that an RDS receiver will cost approximately 10% more than a standard FM stereo receiver. As more RDS receivers are sold and more manufacturers enter the market, the add-on price may drop somewhat.

Why Should I Implement RDS?

For the broadcaster who must make every penny count, this question usually boils down to "how will I make my money back?" Think back to your last station promotion; just how long after the money was spent did the results last? A week? A month? With an investment of about $2500 you can promote your station for a lifetime. Your station will not be promoted on some obscure billboard or bumper sticker, but exactly where you want it—in front of the listener. If you have any doubt that you won't have any listeners with an RDS receiver, rest assured. Every car manufacturer has an RDS receiver waiting for you. All the major after-market providers have an RDS receiver waiting as well. All the home-receiver manufacturers have an RDS receiver waiting for you. Almost every manufacturer has been supplying RDS receivers to Europe for years. The RDS and RBDS systems are nearly identical. Manufacturers want to sell RDS receivers to your listeners, if you will only provide the service.

The next point to make is that radio broadcasters need a technology boost to remain competitive with other media services. The most recent big advent for FM was stereo, and that wasn't last year. The promise of digital radio is years away, and it will cost a lot more money to both the broadcaster and the listener to obtain this service. Any automation system you tie into RDS will be useable on digital radio services, since digital radio will only expand upon the features of RDS. Digital radio will be compatible with RDS, so your equipment will not become outdated or incompatible. RDS provides both you and your listener with the opportunity to become more interactive with technology and with each other. RDS offers the perfect blend of features, performance, and cost that has put it in the mainstream of acceptable practice across Europe.

What Are the RDS Features?

RDS involves the use of primary and secondary features. Primary features are contained in all the information groups, while secondary features are transmitted only if you are using the feature. This prevents the data capacity from being wasted on unused features. The following is a summary of RDS features.

Listener Features

Program Identification (PI) Code. The PI code is a four-digit hexadecimal code that is unique for each station. The PI code is calculated from the station's call letters, thus no two are alike.[1] The receiver uses the PI code to identify your station rather than the frequency. RDS allows frequency diversity so that if you simulcast on another frequency, the receiver can automatically tune to the strongest station. If you do simulcast, then you will pick one of the stations' PI codes and use it on all the stations. While the PI code is transparent to the user, its proper use is vital to proper receiver operation.

Program Service (PS) Name. The PS Name is the name of your station. It is whatever name you choose to present to your listener. It can be your call letters, such as "WZWZ-FM," or a slogan, such as "Z-93." The PS Name is displayed instead of the frequency on an RDS receiver. When the listener tunes to your station, your station name will be shown on the radio display.

The PS Name cannot be over eight characters in length. Any character referenced in the RBDS standard may be used, but special characters may not be displayable on lower-cost British Flag-type displays.[2] The note appearing at the bottom of Table E.1 of the *RBDS Standard* details which characters are most commonly displayed.

Traffic Program (TP). If you offer your listeners traffic bulletins, then pay particular attention to this feature. The TP identifies your station to the listener as one that offers traffic programs. RDS receivers can automatically tune to stations that offer traffic programs when the user turns on the traffic announcement feature of the radio.

Traffic Announcement (TA). When an actual traffic bulletin is broadcast, then this information bit must be set to a logic "1." The RDS receiver detects this and will automatically stop any playback device that may be in use and return to the FM-tuner mode. Audio adjustments are also automatically made so that if, for instance, the user had the volume muted, the receiver will adjust the volume to a user-preset level. Some means must be employed to control the state of this information bit coincident with the beginning and end of the actual traffic announcement.

Program Type Codes (PTY). This code is used to designate the current program material being broadcast. For example, there are predefined codes for "Country," "Rock," and "Top 40." The RDS receiver can automatically tune to stations by PTY, allowing listeners to find their favorite program without tuning to all available stations. Advanced receivers can even interrupt the listener when the "News" is broadcast using a PTY watch mode.

Program Type Name (PTYN). This feature allows the ultimate in flexibility for the broadcaster who desires to be set out from his or her peers. While the PTY codes are predefined, the PTYN can be any eight characters that the broadcaster desires to further describe the current program. For example, a broadcaster who is currently using the PTY "Personality" may set the PTYN to "Limbaugh" to specify the current program. A "Rock" station may set the PTYN to "Bob&Tom" to describe their morning team. An RDS receiver cannot search by PTYN, but will display the PTYN in place of the PTY once it is tuned to a particular station.

Alternate Frequencies (AF). AF allows the RDS receiver to tune automatically to the best station when multiple transmitters or translators are used. Regional or National programs that are broadcast over large areas can even be linked together, providing the listener with the illusion of one very powerful transmitter. It is possible to link stations only during specific times when the program material is common. The RDS receiver will only tune where you tell it to, and only if the PI code is identical.

Radiotext (RT). Radiotext allows you to transmit up to sixty-four characters of information to the listener. Information such as the current artist and song title, station promotional information, local events, and even additional information about the advertiser whose commercial is currently playing can be sent. This is a great tool for interacting with your listeners in a way that was never before possible.

Clock Time and Date. The current clock time and date is transmitted once per minute using this feature. Consumer receivers use this information to keep accurate clock settings.

Emergency Warning System (EWS). The EWS feature allows the transmission of coded emergency information intended for specialized receivers. The FCC recently recognized the capability of this feature, allowing broadcasters to use RBDS within the new Emergency Alert System (EAS). Broadcast EAS equipment can be linked into an RDS encoder for automatic retransmission of EAS information. An alert feature contained within consumer receivers will be activated when an alert code is transmitted.

Data-Related Features

Open Data Channel (ODC). This is the latest and most flexible addition to the RDS standard. In previous years, working groups were formed to determine the best way to add additional data services, such as Differential Correction to the Global Positioning System (DGPS) and even the Emergency Alert System. What these systems had in common was the need for both public (free) and private (fee-based) delivery. Using the few remaining undefined RDS data groups to perform these services proved to be too complex, and coordinated systems could not be agreed upon. All did agree, however, that each data group left undefined was worth its weight in gold, since once all the groups were defined it meant the end of any future expansion to the system.

Thus was born the concept of the open data channel. The open data channel allows the definition of an unused RDS data group based solely on the Application Identification (AID) code. Thousands of AID codes are available, meaning that the remaining data groups could be defined thousands of different ways. The AID codes are internationally assigned and coordinated, so that you (or indeed anyone) can apply for an AID code and start your own data service. The ODC is much more flexible than other data-only groups, such as the transparent data channel, for it can be automatically tuned (by use of the AID) and tracked, regardless of any other data being transmitted. This dynamic, and potentially very profitable, data group will be covered in great detail later in this book.

Transparent Data Channel (TDC). For specialized applications, information of any type can be transmitted by using this data group. For instance, advertising messages could be sent to an electronic billboard using this feature. Virtually any way that you can sell data can be supported by the TDC. Normal car and home receivers do not decode this information.

In-House Application. Data contained in this group is to be used only by the broadcaster. Remote control applications, station telemetry, or paging applications can be supported using this data group.

Radio Paging (RP). Paging services, including those that are numeric or alphanumeric, can be supported utilizing the RDS subcarrier. This is one method for earning revenue from RDS. One company is beginning to build a nationwide paging service at this time and is looking for broadcaster participation.

Traffic Message Channel (TMC). This features allows traffic information to be coded and displayed on consumer receivers so that normal audio bulletins are not required. TMC also utilizes location information such that consumer receivers will display only the traffic information of relevance during a particular trip.

Other Advantages of RDS

The RDS features are quite impressive by themselves. Some other impressive facts that add to the viability of widespread use in the US are:

Over ten years of consumer use throughout Europe

Low-cost receivers offered by almost every major manufacturer
    Automotive
    Home
    Portable
    Personal Computer
    Pocket Pagers
    Data-Only Receivers

Low investment cost of equipment to the broadcaster
    RDS Encoders
    Modulation/Data Monitors
    Automation Equipment

Potential revenue to the broadcaster
    Increased listener interaction with the station
    Possibility to lease part of the RDS subcarrier to a third party
     providing data services
    Paging and other data services
    Electronic Billboards

Ability to link simulcast transmitters as "one" through use of the alternate frequencies feature
Potential to share services through use of the Enhanced Other Networks feature
The following chapters will provide further detail of each of these features, as well as detailed descriptions of available products. Detailed implementation instructions will be provided so that your trip into RDS will be quick and easy.

The Importance of Standardization

The fact that there is a standard for RDS means quite a bit. The competition of rival technologies often results in lack of participation by manufacturers and broadcasters alike. The AM stereo standard was one such debacle. When the industry could not arrive at a consensus, the FCC was drawn in to make the decision. Nearly ten years later, the FCC finally did reach a decision. However, in that time-frame, broadcasters failed to make the commitment necessary to convince receiver manufacturers to include the feature in products. Consequently, by the time the standard was approved, there was little industry support remaining. The attempt to revive AM stereo after the standard was approved might be likened to kicking a dead horse to see how far it will run.

While standardization in and of itself does little to guarantee the success of a technology, it does provide one of the key factors required that can allow a technology to grow. The RBDS standard provides the framework around which an entire industry can be built. While the particular features of a receiver or an encoder may be different, they are all designed to meet the basic criteria of the protocol. This standardization framework provides the structure that will support the use of RDS throughout North America. A standard is like a set of building codes that describe the requirements of the framework, yet still allow each house to have its own unique look and features.

Standards in the United States do not come about easily. Anyone who has ever participated in a standards organization will tell you that that comment is a gross understatement. The RBDS standard provides the coordination that allows some 6,000 broadcasters and dozens of equipment manufacturers to provide a consistent and uniform technology to listeners. Imagine trying to provide a service such as this without the use of a standard.

Notes

1. See Annex D of the *RBDS Standard.*
2. See Annex E of the *RBDS Standard.*

General Managers' Overview

RDS is the Radio Data System. RDS is the name of the European standard as well. In the U.S., the standard is called the Radio Broadcast Data System, or RBDS, but for your listener, the technology is always referred to as RDS. RDS is a method of transmitting program-related data inaudibly on a 57 kHz subcarrier. The following question and answer section will answer the questions most important to you. Abbreviations particular to RDS are shown in parentheses for reference.

Frequently Asked Questions

Q: How much does it cost?

A: An RDS encoder may be purchased for as little as $1000, or as much as $4000. Before purchasing an encoder, determine how you want to use RDS at your station. You may be better off buying a more expensive model with more features now rather than scrapping a model later that does not meet your long-term requirements.

Q: Do I need a dedicated computer for RDS?

A: While some units may be ordered preprogrammed with fixed station information, many require at least part-time use of a personal computer in order to program and update information in the encoder. To take full advantage of all RDS has to offer, you might consider dedicating a computer for RDS use or purchasing automation software that will automatically update program information. (This may be a good time to upgrade your own system and give that old model over to the engineer!)

Q: Does RDS require a special receiver ?

A: Yes. In order to receive RDS data, you must purchase an RDS capable receiver. Receivers are available for the car, home, portable, and personal computer. Figure an additional $25 to $50 cost for an RDS radio, depending on the features. An RDS receiver will be marked with the RDS logo on the front.

Q: Who makes RDS receivers?

A: Almost every major manufacturer. There are currently over 400 models available in Europe. While there aren't as many available in the U.S., that will change rapidly. Home and after-market car receivers are currently available from Denon, Kenwood, Pioneer, Onkyo, Bang & Olufson, and Advanced Digital Systems (computer receiver). RDS is standard in all Porsche and Audi A4 vehicles. General Motors began shipping RDS as standard in upper-level car receivers in some models this year. Expect Ford and Chrysler, as well as other major manufacturers, to follow GM's lead.

Q: Can I buy an RDS converter to add RDS to my existing receiver?

A: Not really. Many RDS features are highly integrated into other receiver functions, such as the seek, scan, clock, and volume. Other RDS features require integration with the tuner and display, so an "RDS converter" would be limited in features and functionality. Such a device would be limited to the display of only station information.

Q: If I use RDS can I keep my existing subcarriers?

A: In most cases, yes. One subcarrier service, Cue Paging Corporation, utilizes the 57 kHz subcarrier as well. It is possible for RDS and Cue Paging to coexist, utilizing a time multiplexing format. This format is detailed in Annex P of the *RBDS Standard*. If you currently lease your subcarrier to Cue, you must renegotiate your contract before transmitting RDS data. There will most likely be a cost penalty involved.

Q: What type of information can be sent?

A: The most popular features of RDS are related to the current program being aired. RDS will give you the following capabilities:

1. *Display the name of your station.* The Program Service (PS) Name allows you to send your station's call letters or slogan, such as "WDFM-FM" or "Rock-104." This information will replace the frequency display

on an RDS receiver, giving instant recognition of your station to anyone tuning to your frequency. Up to eight alphanumeric characters may be defined for your station's PS.

2. *Broadcast the program content.* The Program Type (PTY) feature allows you to tell the listener what the current program material is, such as "Classical," "Rock," or "Top 40." Listeners may search for their favorite program type. They may also select a program type such as "Weather" or "News"; when it becomes available, it will interrupt their cassette or CD. Using the Program Type Name (PTYN) feature you may give further detail (up to eight characters) about the current program. Some examples are "Altrntve" for alternative rock or "H. Stern" for Howard Stern. This information is automatically displayed when anyone tunes to your station.

3. *Traffic Information.* As in the case of the program-type feature, you may interrupt a listener's cassette or CD when a traffic bulletin is given. Listeners may search for stations that provide traffic bulletins.

4. *Free format-text messages.* Using the radiotext (RT) feature you may transmit messages of up to sixty-four characters to your listener. Some examples include, current artist and song title, next up, station promotions, or advertiser information during commercials, such as store hours, location, or phone number.

5. *Broadcast Emergency Bulletins.* In compliance with the FCC's Emergency Alert System (EAS), RDS may be used to alert listeners (even with the radio turned off!) to emergency messages.

Q: Are there any datacasting opportunities with RDS?

A: Yes. RDS provides a multitude of ways to transmit data for yourself or others willing to lease subcarrier capacity. Some examples are paging and differential corrections to the global positioning system (GPS). Since RDS is limited to 1200 baud capacity, you will want to carefully review capacity requirements before signing any contracts. RDS also has a powerful feature that allows you to define data in an open format, giving you the opportunity to create your own unique data service.

Q: Will RDS be replaced by new technologies, such as high-speed subcarriers or digital audio broadcasting?

A: Not for a long time. RDS has been adopted as a standard by the National Radio Systems Committee (NRSC). The NRSC is a joint committee of the National Association of Broadcasters and the Electronic Industries Association. No other data transmission scheme has approval for use by the NRSC. All future systems will have to be compatible with RDS. Eventually, when a Digital Audio Standard (DAB) is adopted, RDS may be replaced along with your analog audio. All the automation concepts and features of RDS will be duplicated and expanded upon with DAB, making your investment in RDS a wise choice.

Q: Will RDS improve my audio quality or transmit range?

A: In and of itself, no. But the Alternative Frequency (AF) feature will allow you to increase the apparent power of your station to the listener. The AF feature allows an RDS receiver to automatically switch between multiple transmitters broadcasting on different frequencies. Using translators or multiple lower-power transmitters along with the AF feature will allow you to increase your listenership without the investment of a larger transmitter or frequency move for your existing station. Since the listener only sees the name of your station (with the PS feature), the currently tuned frequency becomes irrelevant.

Q: Won't the use of RDS allow other broadcasters to "steal" my listeners?

A: No. Each and every station has a unique Program Identification (PI) code that is used by the RDS receiver to ensure that it remains tuned to the desired station. Every time the listener recalls a radio preset, or an AF switch is performed, the stored PI code is verified before the audio is unmuted. The interruption of the cassette or CD for traffic, weather, or news programs only occurs for the station that the receiver is tuned to.

Q: Can RDS increase my ratings?

A: Quite possibly. Listeners will tune to your station to obtain the services their new RDS receiver offers. By using the RDS features to their fullest, you may attract new listeners who are interested in this new era of "interactive radio." By using the alternate frequency feature, you increase your listenership by sheer coverage area and keep people tuned to your station without requiring the listener to remember multiple frequencies or retune the receiver.

Q: Isn't RDS new and unproven?

A: No. RDS has been in use throughout Europe for more than twelve years. Over 35 million home, car, personal computer, and portable RDS receivers have been sold. Every major manufacturer has a wide variety of product offerings available. With over 600 stations now broadcasting RDS in the U.S., RDS receivers will be hitting the market in large quantities.

RDS Is Different from Any Other Subcarrier

RDS is different in that the information you transmit is sent directly to the listener. RDS changes the way you can interact with that listener. Other subcarriers are often leased out to a third party in exchange for some sort of monetary compensation. Aside from the initial contract negotiation, and periodic monitoring of injection level, the broadcaster involvement is none. RDS completely reverses both your role and involvement with your subcarrier service. RDS is your subcarrier and your new link to your most important customer; your listener. Obviously, some thought, planning, and activity must surround this new service for it to be truly successful. You will be able to motivate an entire industry to supply products and services that will allow you to take every advantage of RDS.

The most important message that you can learn is that it is not business as usual. Something really revolutionary is at your fingertips, something that will change the way you do business. The interaction that is now possible is the key to continued success. The competition for advertising dollars by other media is staggering. The broadcaster that takes advantage of the benefits of RDS will be able to attract more listeners and exchange information in ways never before possible. Out of all this it is you, the broadcaster, who stands to benefit the most.

Do I Have to Give Up an Existing Subcarrier to Transmit RDS?

The only other existing service that resides on 57 kHz is a paging service that is operated by Cue Paging Corporation. This subcarrier is actually the predecessor to RDS and is known technically as Mobile Search or MBS.[1] This paging system was developed in Sweden; it was eventually phased out in Europe in favor of RDS paging services.

MBS shares the same transmission method as RDS, as well as the same basic data structure, but the actual data protocols are different. RDS essentially adds more potential features and services (including paging), but at the expense of paging capacity.

The *RBDS Standard* provides the capability to time multiplex RDS with MBS. This protocol is referred to as MMBS.[2] While it is technically possible to transmit MMBS, you must obtain written approval by the paging provider before you can do so. Generally, a written contract prohibits such operation, so you must specifically request approval to transmit MMBS at the time of contract negotiation. It is important that you obtain in writing the exact number of RDS data groups that can be transmitted in a given time interval. Since transmission of MMBS will reduce the maximum paging capacity of the service, you typically will earn a proportionately decreased amount of revenue from the paging provider.

It should also be pointed out that transmission of MMBS may cause degradation of receiver performance due to the reduced repetition rate of RDS information. For instance, an RDS receiver that is performing a Program

Type (PTY) search, may miss an MMBS station, because it is unable to synchronize and obtain the transmitted PTY within the designated seek time. Alternate Frequency (AF) switching may be slowed or sometimes entirely corrupted when tuned to an MMBS station. Chapter 17, on RDS group structure, will explain the tradeoffs in further detail.

Aside from MBS paging services, there are no other subcarriers in existence on 57 kHz. If you are currently transmitting other subcarriers, then RDS will cause little or no interference with them; likewise, they will cause little or no degradation to RDS. Subcarrier compatibility is covered extensively in chapter 4.

If other subcarriers are utilized along with RDS, you must ensure that the total maximum deviation of all subcarriers is limited to 10% (referenced to 75 kHz).[3] RDS typically requires an injection level of about 3.75%, leaving 6.25% for other subcarrier services. As in the case of MBS, if these other subcarriers are leased out, the contract should be carefully reviewed for the minimum allowable injection level. RDS injection levels can often be run as low as 1.6% with little degradation, but actual field measurements should be obtained before decreasing the injection level below 3.75%.[4]

Conclusions

RDS offers you the opportunity to interact with your listener by providing important information about the programs and services you offer. You may attract new listeners or keep long-time listeners more interested and involved. With a minimal investment, you can take advantage of a proven technology that will carry you into the future.

Notes

1. Swedish Telecommunications Administration (Televerket), *Paging Receiver for the Swedish Public Radio Paging System*, Specification 76-1650-ZE (1976).
2. See Appendix K of the *RBDS Standard*.
3. See Chapter 5, "FCC Rules Concerning Subcarrier Usage."
4. See Chapter 19, "Injection Adjustment."

3

Summary of Changes in the New Standard

The newly revised *RDS* and *RBDS Standards* move forward into a new era for FM data services. All changes have been devised to be fully compatible with the old standard, while offering a tremendous amount of growth. The prior standards were:

- *United States RBDS Standard—January 8, 1993*. Issued by the National Radio Systems Committee (NRSC).
- *EN50067, Specification of the Radio Data system (RDS)—April 1992*. Issued by the European Committee for Electrotechnical Standardization (CENELEC).

While these standards shared much commonality, the newly revised standards are almost completely identical in form. In this chapter, we will explore the changes that have taken place and also take a close look at the remaining differences between the new standards. The following is a combined list of changes to both standards with detailed explanations of why these changes have been made. The list of changes are as follows:

- Program Type Name (PTYN)
- Fast Program Service Name (Fast PS)
- Location and Navigation (LN)
- Open Data Applications (ODA)
- Enhanced Radio-Paging Protocol (EPP)
- Language Identification (LI)
- Decoder Information (DI)

- Extended Country Code (ECC)
- PI Structure from B000-FFFF
- Analog SCA Cross-Referencing
- IRDS Updating
- PTY Code Table
- MBS EAS (Emergency Alert System)

Program Type Name (PTYN)

The PTYN contained within group 10A has been added to the *RDS Standard*. This feature was previously defined only within the *RBDS Standard*. The feature has been slightly modified with the addition of a text A/B Flag, but is still compatible with the previous definition. The A/B flag is located in the unused four-bit region and shares the same bit position as the A/B flag in Radiotext group 2A. The usage is the same in both groups in that the A/B flag should be toggled whenever the PTYN is changed. Within the receiver, the PTYN text buffer should be cleared when the A/B flag is toggled to prevent "mixing" of old and new PTYNs. The rationale is that the PTYN is designed to be semidynamic, potentially changing whenever the program changes. Newly designed receivers can take advantage of the A/B flag without effects from existing encoders that do not incorporate this change.

Fast Program Service (PS) Name

The Fast PS is currently defined in group type 15A of the *RBDS Standard* and is not defined in the *RDS Standard*. This group will not be adopted in Europe. In North America there will be a phase-out of all transmissions within ten years, and within two years for new encoders. The elimination of this group can be included in encoder upgrades to the new *Standard*. This group will not be defined for any use and could later be reallocated as open data. This will not adversely affect the existing receiver base, for only Delco Electronics, Grundig, Blaupunkt, and Becker have produced receivers for North America with this feature. The rationale for this change is that it will not cause significant performance degradation in the receivers because the PS is also located in groups 0A and 0B. The PS is transmitted two characters at a time in these groups and will be received in 348 ms, versus 174 ms using the fast PS feature.

Since the PS is typically static and is often stored in receiver memory for instant recall following PI reception, the only time that slower PS acquisition will be noticed is when tuning to a station for the first time. There are over 35 million receivers in use worldwide that would not be able to receive the fast PS. Once broadcasters would begin use of Fast PS, acquisition times of PS in existing receivers would be adversely impacted since they would not be able to decode this data. Therefore, as a feature, Fast PS cannot be considered fully backwards compatible. The fast PS would be of minor benefit, especially in Europe where most broadcasts are network-based and hence require frequent repetition of AF information (0A). So, in consideration for phase-out in North America, the Europeans have agreed to coordinate the future use of 15A.

Location and Navigation (LN)

LN was defined in group 3A in the *RBDS Standard* only. This usage will be immediately deleted and reassigned to the Open Data Application (ODA). This was recommended since there were no LN transmissions within North America and the feature was no longer required for ID Logic receivers. The definition of LN was found to be inadequate as determined by the Differential Global Positioning System (DGPS) working group findings. This group, working along with a similar European group, determined that the LN structure cannot meet the current needs of private and public service providers. It was recommended that the Open Data Application be used for LN and DGPS services. The Immediate reassignment of group 3A from LN to ODA will yield a global ODA structure with internationally allocated application identification codes.

Open Data Applications (ODA)

The Open Data Applications (ODA) has been newly defined for group 3A, which required reassignment of location and navigation (LN). The use of an "A" type group allows for the low-data capacity applications to exist entirely within group 3A. The ODA allows multiple reuse of unused data groups rather than fixed definitions, ensuring a long future for RDS. The actual use of data can be either public or private. The ODA allows encryption of data for fee-based services. Data groups referenced via slow labeling (group 1A) can also be used for ODA when not specifically referenced for use. Specific uses are identified by application identification codes (AID). Over 65,000 AID Codes are available and are internationally allocated. This allows anyone to start their own datacasting business locally, nationally, or internationally. Open Data Applications are already defined or proposed for use for EAS, DGPS, DAB Cross-referencing, and enhanced TMC.

Enhanced Radio-Paging Protocol (EPP)

To enhance the existing paging capabilities of RDS, the enhanced radio-paging protocol was developed. This protocol is contained within groups 13A and 1A. Annex M of the *Standard* contains the protocol definitions. With the enhanced protocol, the possibility of international RDS paging has been added. The prior definition of RDS paging was based upon the usage of PI codes only. This worked well in network-based broadcasts, as found in many parts of Europe, but offered no solution where non-network-based broadcasts are employed, such as in the U.S. The enhanced radio-paging protocol now opens opportunities for local, national, and international paging opportunities.

Language Identification (LI)

To aid the international traveler, the language identification feature located in group 1A, Variant 3, has been expanded to include almost all world languages. The list of covered languages is located in Annex J of the *Standard*. The feature enables the broadcaster to indicate the spoken language he is currently transmitting, thus allowing receivers to search for their native language on local broadcasts.

Decoder Identification (DI)

The decoder information feature has been modified to introduce a new feature. The DI is located in type 0A, 0B, and 15B groups. Its usage is detailed in section 3.2.1.5. This enhancement maintains backwards compatibility through the use of single-bit identification, rather than a look-up table, allowing the additional new feature for identification of dynamic PTY switching. This allows the broadcaster to identify that he dynamically switches the program type (PTY) based on the current audio program, thus supporting the use of PTY Interrupts in consumer receivers. Prior to this, there was no way to tell if the broadcaster would ever change the PTY. The dynamic PTY indicator is analogous in use to the TP bit of the traffic announcement feature.

Extended Country Codes (ECC)

The ECC codes allow receivers to identify the country that the broadcast is coming from. Since PI codes are limited in number, they must be repeated throughout the world; when the PI code is received in conjunction with the ECC, the exact country of origin can be identified. The updated ECC code table has been expanded to be international in scope. It is contained in Annex D, and Annex N. The ECC should be transmitted by all broadcasters, and it is recommended that it be a default automatic transmission in encoders. The ECC supports the development of global receivers that can automatically compensate for things such as:

- E blocks
- PI codes
- PTY tables
- Tuning range and step
- FM deemphasis

Program Identification (PI) Codes

In the U.S., PI codes are based on call letters rather than being assigned by any organization as is done in Europe. A portion of the PI codes are reserved for network usage and also for assignment to stations in Canada and Mexico. During the upgrade to the *Standard*, several mistakes were discovered in the non-call-based PI codes that had to be corrected. The changes to the PI code assignment are summarized as follows:

- PI assignments below B000 will remain as they are, allowing AF switching but no regionalization. PI codes "_0_", and "__00" are remapped into the "A" range of PIs.
- C000 to CFFF are assigned to Canada. This allows AF switching, but no regionalization. PI codes C0xx, and Cx00 are excluded from use.
- F000 to FFFF is assigned to Mexico. This allows AF switching, but no regionalization. PI codes F0xx , and Fx00 are excluded from use.
- B_FF and D_01 to E_FF are assigned for national networks in the U.S., Canada, and Mexico. Regionalization is allowed. NRSC will provide assignments for all three countries.

The ECC code table was modified for these changes as well. The rationale for these changes is that the current definitions for PIs above BFFF are contradictory. Also, the present PI code structure above BFFF was inadequate for the needs of Canada and Mexico, because the existing PI structure only allowed 256 PIs for Canada and Mexico. The new PI assignments yield 3,584 possible nonregional PIs for Canada and Mexico, as well as 765 national network PIs for all three countries.

Analog SCA Cross-Referencing

This feature was devised to allow RDS signaling to receivers equipped with analog subcarrier decoders to switch over automatically. This feature was defined in section 3.1.3.6, note 3 as part of the Transparent Data Channel (TDC), group 5A, channel 2. This feature will be deleted from the *Standard* for the following reasons: it is not in use; it prevents harmonization with the *RDS Standard*; and it can be easily be converted over to the ODA channel.

ID Logic RDS (IRDS) Updating

This feature was previously defined in section 3.1.3.6 notes 1 and 2 as a reservation of the Transparent Data Channel (TDC), group 5A channel 0 and 1. This was reserved for updating an ID Logic receiver database through RDS. It was agreed that the TDC reference would be deleted from the *Standard*. The rationale behind this change is that it is not currently in use and prevents harmonization. The owners of the ID logic patent, PRS, have agreed to move the IRDS updates to the ODA. The use of AID codes will provide better protection and international usage of the IRDS feature.

PTY Code Table

Both the RDS and RBDS standards have included additional definitions for previously undefined PTY codes. RBDS has defined two new codes:

- PTY-29, Weather. It is defined for nonemergency weather-related information, such as forecasts, watches, and advisories. Weather is a popular listener feature that is constant. It can be supported with the PTY "Watch" mode to interrupt playback, in a manner similar to traffic announcements and News PTY.
- PTY-23, College. It is intended to identify broadcasts oriented to college students. These broadcasts often contain a variety of formats and information. "College" is one of the top three noncommercial formats. The other two, "Public" and "Religious," already have PTY definitions.

In the *RDS Standard*, all the PTY codes have now been given definitions. The main purpose of this was to ensure coordination with the development of the Digital Audio Broadcasting system, known as the Eureka 147 Standard, being rolled out throughout Europe. Previously, codes 16 through 30 had no definition. These codes have all been defined in the new *Standard*.

Differences between RDS and RBDS

With all the changes to the two standards, there still remain differences. These differences are mainly market driven. There will always be differences in the actual broadcast and market infrastructure that cannot be taken into account without employing modifications. From the listener's point of view, most of these differences are transparent. In exploring these differences, it will become evident that the receiver manufacturer must be well-versed in the differences to ensure proper receiver operation. When these differences are properly employed by the receiver manufacturer, it is possible to construct a truly global receiver. If broadcasters and receivers make use of the ECC feature, it is possible that the receivers could self-configure automatically. The problem, of course, is getting broadcasters to utilize the ECC feature, since it is not a required broadcast feature.

Summary of Differences between RDS and RBDS

- *Program Type Definitions (PTY)*. To conform with differing broadcast styles, the PTY code definitions are different. These differences may be accounted for through the use of a look-up table within the receiver. Annex F of the *RBDS Standard* includes both the European and North American PTY definitions.
- *Program Identification Coding (PI)*. Due to the high penetration of network-based programming throughout Europe, as compared to the largely independent single transmitter structure of North America, the derivation of PI codes is different. In the *RBDS Standard*, PI codes are based on call letters from 0000 to AFFF and are network-based from B000 to FFFF. This means that PI codes above B000 in North America are treated by the receiver the same way that all PI codes are treated in Europe. Below B000, PI codes in North America do not employ the regionalization feature. While alternate frequency switching is still employed below B000, variants based upon changes in the second nibble of the PI code do not exist and should be ignored. PI code definitions can be found in Annex D of the *Standard*.
- *Mobile Broadcast System (MBS) and RDS/MBS Time Multiplexing (MMBS)*. The predecessor to RDS, MBS, still thrives throughout North America. MBS is mainly used as a paging system through a network of approximately 500 stations within the U.S. The MBS system utilizes the same modulation and data structure as RDS but employs a different data protocol. An MBS broadcast is identified through the offset word E. Since there are similarities between the two systems, it is possible to time multiplex MBS and RDS data. This time sharing is known as MMBS. Receiver manufacturers must be able to differentiate between RDS and MBS, as well as accommodate MMBS broadcasts. Internal MBS/MMBS cross-references can be found throughout the new *RBDS Standard* as a reminder of particular system requirements and as a possible alternative to RDS. A public domain EAS protocol is also contained within the MBS/MMBS annex of the *RBDS Standard*.
- *Emergency Alert System (EAS) ODA Protocol*. Within the *RBDS Standard*, the EAS ODA protocol is defined for use in the United States. This optional feature set is constructed around the FCC's newly de-

veloped EAS system and is open for public use. RDS allows the silent retransmission of emergency information. This has been combined with existing consumer-oriented emergency features to allow additional feature functionality for consumer receivers as well. The EAS ODA can also accommodate private emergency systems.

- *ID Logic Feature (IDL)/In-Receiver Database Updates via RDS (IRDS) ODA Protocol.* The ID logic feature is a licensed technology that allows the incorporation of an in-receiver database that contains format type and call letters for all AM and FM stations. When combined with RDS, IDL can provided similar data and features for non-RDS and AM stations. The IRDS feature allows the database to be updated through an RDS ODA so that the information can be updated and maintained automatically. The introduction of ID logic is contained within a separate section of the *RBDS Standard*, while the IRDS ODA is described in detail within an annex of the *Standard*.
- *AM RDS (Future System).* Although a suitable data transmission system has yet to be developed that is compatible with the AM stereo system (CQUAM), there is a separate section where such a system can be defined or referenced.

4

RDS versus Competing Subcarrier Technologies

Subcarrier technologies are progressing at a phenomenal rate. RDS has been in use for over ten years to date. So what makes RDS worthwhile to implement? Why bother at all with an antiquated subcarrier system when so many faster alternatives are available. The answer is that all the other systems you hear about are not really available. For a subcarrier system to be truly available for use it must meet the following requirements:

1. *It must be standardized.* RDS is the only system to date with approval of the NRSC. This approval means that both broadcasters and consumer manufacturers support the system. Since the FCC deregulated subcarrier use, you really can put anything into the subcarrier that meets certain minimum FCC criteria. RDS is oriented to your most valuable asset—your listener. Don't waste your time on anything that doesn't have a standard to support common implementation, unless all other criteria in this list is met, making it a de facto standard.
2. *Low-cost implementation.* To reach a broad base of listeners, your subcarrier system must be inexpensive to implement. Manufacturers support RDS because of the wide availability of low-cost integrated circuits, as well as its minimal impact to other system components. You can purchase an encoder for under $1000.
3. *No interference to the main program.* For all of you out there actually considering using In Band On Channel Digital Audio Broadcast (IBOC DAB) technology, wake up. Talk to your engineer about subcarrier interference to your main audio channel. The modulation

method and low injection rate of the RDS subcarrier will cause the least amount of interference to your main audio channel in comparison to anything else. IBOC DAB is a thousand times worse than any subcarrier data system. Subcarrier interference is worse in areas of multipath interference, where mixing products between the subcarrier and main audio cause disturbances in the recovered audio output.

4. *It should be robust.* Mobile reception is the worst application for any system. Fading, multipath, and RF interference can cause great harm to the recovered data. RDS was designed for robust mobile reception. It is slow, but it was designed from the ground up to be mobile compatible.

5. *It should be compatible with existing services.* During laboratory tests of In Band Digital Audio Broadcast and high-speed data subcarrier systems, RDS has been proven to be compatible with all existing and proposed systems. You can implement RDS where you already have two existing analog subcarriers. All newly proposed systems would force you to eliminate at least one, if not both, of your existing services. In many cases these existing services bring in revenue for your station. RDS allows you to keep your existing services.

6. *It should be proven.* RDS has a track record of over ten years in Europe. There are over 500 models of RDS receivers available today. RDS has been accepted by the consumers, broadcasters, and manufacturers. As a system, RDS offers value-added data to listeners about the broadcast, without deluging them with too much information. This makes the system compatible in all environments, such as home, portable, and automotive, with minimal cost added.

The RDS Subcarrier

Figure 4.1 depicts the audio spectrum of an FM broadcast station. RDS nests into the 57 kHz position between the stereo multiplex and the 67 and 92 kHz subcarrier channels. Injection levels as low as only 1.3% allow RDS to be easily implemented without giving up program audio power. The amplitude modulated, double sideband, suppressed carrier (AM DSB SC) modulation causes the least amount of possible interference to the main audio channel. The RDS subcarrier is phase-locked to the third overtone of the pilot signal, further minimizing possible interference.

Leasing Your RDS Subcarrier

Since RDS is such an attractive system for robust data transmission, it is no surprise that many companies have developed data services utilizing this subcarrier. In fact, with the development of the Open Data Application (ODA),

Stereo Multiplex Signal with RDS and 2 SCA's

Monaural Signal L + R	Pilot tone	Stereo Signal L - R	RDS Signal	SCA Signal	SCA Signal

15   19   23   53   57   67   92   kHz

Figure 4–1   A typical broadcast system.

anyone can begin an RDS data service. You might be approached about leasing your RDS subcarrier in the near future, so knowing some basics will help you make an intelligent, informed decision that offers you the best of both worlds—listener services and additional revenue. When negotiating a lease contract, you should have an idea of your long-term usage of listener services, since the more listener services you offer, the less capacity you will have remaining for non-listener-oriented data. Chapter 18 discusses RDS features and capacity requirements in detail, but for now, answer the following questions before considering a sublet of the RDS subcarrier, as it is possible to lease part of your RDS capacity:

1. What consumer oriented features will you offer? The most important RDS features are contained in multiple groups, reducing the impact of a sublease. What will your needs be? For instance, will you implement radiotext or the program type name feature?
2. What is the total capacity required to do what you want to do? How much is being offered? You must carefully weigh your needs against what is being offered. Will I attract or lose more listeners if I can't utilize radiotext? The station engineer should carefully review the required transmission requirements for each feature you would like to implement to ensure that they can be correctly implemented within the sublease agreement. The RDS group structure should be thoroughly analyzed.
3. Who owns, operates, and maintains the equipment? Can you dynamically change your listener-oriented services, such as the program type code? What are the interface and communication protocols of the system? Will you be supplied with interface control software and hardware?
4. How much injection is required? Is the required injection within the FCC guidelines when your other subcarriers are taken into account?

5. How long is the lease? If you sign the lease agreement and are not satisfied, can you break the contract without severe penalty?
6. How much will you earn? Is it a fixed amount or dependent upon subscribers?

MBS/MMBS Paging System

The predecessor to RDS is known as the mobile broadcast system or MBS. This system operates on the 57 kHz subcarrier and employs the same modulation method and block structure as RDS but utilizes a different group structure. In Europe, this system was phased out, then implemented within the RDS paging protocol. In North America, however, the MBS system still exists on over 500 stations. The *RBDS Standard* makes provision for the MBS and RBDS data to be time-multiplexed, thus allowing both systems to coexist. This application is referred to as MMBS and is depicted in Figure 4.2. As you can imagine, whenever two different data protocols are time-multiplexed, the data rate of either system must be reduced since the total of the two cannot exceed 1187.5 baud. If you already have or are considering a lease agreement involving this system, you must specifically work out the time-sharing arrangement in the contract. The questions outlined above apply to MMBS just as they would to subleasing part of your data capacity for an RDS-based paging application, with one exception. Since the MBS and RDS systems have a different data protocol, no RDS data is carried during the time period that the MBS data is being transmitted. This means that the program identification codes, program type codes, traffic announcement flags, and so on are not transmitted during the MBS time period. This does cause degradation of consumer-receiver performance, for it attempts to tune to your station. Receiver manufacturers can make design provisions for the MMBS protocol, but there will always be a loss of RDS data during the MBS time period.

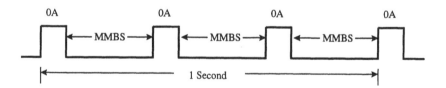

Figure 4–2 MMBS: The time-multiplexing of RBDS with MBS.

Other Factors Involving MMBS

The issue of reduced RDS information rate (and reduced responsiveness) due to its being displaced by paging information is easily understood. The rate reduction causes the most severe problems where alternate frequencies are involved and may be perceived as a problem with PTY search. The synchronization method of MBS also differs from RDS. MBS packets are of variable lengths with identifying header blocks. The systems differ in the way the starting point of a group or packet is determined. MBS takes advantage of known header information to locate the start of a relevant packet with a high degree of confidence. RDS, which must accommodate arbitrary group content, relies on identifying the CRC fields. Individual RDS blocks (and thus groups) are identified by adding a specific offset to the CRC of each block. The offset constants are called A, B, C, and D, respectively. As a special case, block 3 has an alternate offset, C', to allow decoding one particular meaning without first correctly decoding block 2. The offsets do not affect the error-detecting and correcting capabilities of the CRC, but are chosen such that a shifted code word (a legitimate twenty-six bit string displaced one notch) will not be near enough to another code word and be mistaken. The MBS system does not offset the CRC (the zero offset is called E) and a shifted code word has a 25% chance of matching some other code word.

The real concerns have to do with system robustness in the presence of noise. In a low-noise environment, the two data formats can peacefully coexist on the same channel without ambiguity. In a noisy environment, the situation becomes more confused. The problem is to properly synchronize to each type of block. The choice of zero for the E offset is particularly unfortunate because a one-bit shift has a significant probability of mimicking a legitimate block. The possibility of randomly inserting a random number of MBS blocks between RDS groups destroys the fixed progression of offsets and makes "flywheeling" through noise bursts more difficult. With more offsets to choose from, there is more opportunity for something to go wrong. It is clear that multiplexing RDS and MBS data will cause some degradation in noise tolerance. The magnitude of the degradation, or even how to quantify it, is not clear. Effects over narrow ranges of conditions need to be considered, because some users spend a large portion of their time in fringe areas.

5

Regulation of the RBDS Standard

The *RBDS Standard* is maintained by the National Radio Systems Committee (NRSC). The NRSC is a joint committee of the National Association of Broadcasters (NAB) and the Electronic Industries Association (EIA). An NRSC standard is, in essence, an agreed upon standard between the broadcasters and the equipment manufacturers. While an NRSC standard is a voluntary standard, as there are no FCC rules governing subcarrier usage to this degree, it is an agreement to adhere to a set of written rules or guidelines to insure interoperability and support of certain features. Without this standard there would be no assurance that a given receiver would work the same way from station to station. The *RBDS Standard* covers all of North America, including Canada and Mexico. The newly updated *Standard* is essentially a global standard in which all the same features and protocol are applied. The result is that equipment manufacturers and service providers can now supply one model that will work globally.

The NRSC RBDS Subcommittee

The RBDS Subcommittee was formed out of the NRSC full committee to administer the RBDS specification directly. This committee is responsible for writing the actual *Standard* and updates as decided upon by the subcommittee. All recommendations made by the subcommittee are then passed on to the full NRSC committee for approval if incorporated as part of the *Standard*. The RBDS subcommittee was responsible for formulating all modifications to the European *RDS Standard* for use in North America. They continue to work closely with the European RDS forum to ensure continued support and education of interested parties. Since the two standards are so closely related, many support documents produced in Europe can be distributed for use in North America with little or no modification.

The *RDS Standard*

The European *RDS Standard* is formally issued by the European Committee for Electrotechnical Standardization (CENELEC). The *RDS Standard* is issued under document number EN 50067 and was last revised in April 1992. The new *Standard* will be adopted sometime in 1997. While the CENELEC committee is formally in charge of the *Standard*, a voluntary group known as the RDS Forum is the actual mechanism that drives changes in the *Standard*.

The RDS Forum

The RDS Forum is organized under the European Broadcasting Union (EBU). The RDS Forum includes broadcasters, receiver manufacturers, and governmental agencies among its members. What makes the RDS Forum so effective is that its members are all long-time users of the RDS system who understand the distinct differences between the real world and the paper standard. Besides working directly with standards issues, the RDS Forum publishes many informational documents that help support proper usage of the *RDS Standard*. Membership to the RDS Forum is open to any organization and costs around $1100 per year.

Consumer Electronic Manufacturers Association (CEMA)

CEMA is a branch of the Electronic Industry Association (EIA) and, as the name implies, focuses its efforts on manufacturers of consumer equipment. In an unparalleled effort, CEMA cosponsored an RDS promotional effort in 1995-1996 that offered RDS encoders and receivers to broadcasters in exchange for advertising. The ads were aimed at informing and educating consumers about RDS. In addition to education, the promotion helped to break the "chicken-and-egg" situation that existed with RDS: broadcasters were reluctant to purchase encoders due to the lack of widespread availability of consumer receivers, and receiver manufacturers were reluctant to produce receivers with so few broadcasters on the air with RDS. The campaign involved Delco Electronics, Denon, and Pioneer, who provided funds that were matched by CEMA. This campaign focused on the top fifteen radio markets in the U.S. and was highly successful. Not only did the campaign add over 350 RDS stations, but it also helped convince many manufacturers that it was time to produce receivers for the market.

The RDS Advisory Group

In North America, the RDS Advisory Group serves as the counterpart to the European RDS Forum. Since most of the technical development of the RBDS and RDS standards are handled by other organizations, the RDS Advisory group focuses on educational and marketing issues. The Advisory group sponsors joint efforts, such as a consumer educational brochure on RDS features and benefits, as well as a quarterly newsletter covering recent RDS developments and news. By pooling resources, the group is able to accomplish much more than any individual company can alone. The Advisory group is beneficial to manufacturers, broadcasters, and service providers alike. Membership in the RDS Advisory group is open to anyone at no charge. The Advisory group is sponsored by CEMA.

FCC Rules Concerning Subcarrier Usage

The FCC has deregulated subcarrier usage, allowing almost any technology to be used as long as certain criteria are met. This means that analog or digital or a mixture of both may exist within the subcarrier frequency range, as long as the total modulation does not exceed 110%. RDS has proven to be very compatible with existing services due to its low injection requirements and narrow bandwidth. In addition, by phase-locking the AM double sideband suppressed carrier signal with the stereo pilot signal, there is very little risk of interference to the main audio program, even under multipath conditions. RDS is carried on many classical networks throughout Europe and the U.S. with no degradation to the audio quality.

Learning More

Information about RDS can be obtained from any of the organizations discussed above, including:

1. The National Association of Broadcasters (NAB).
2. The Consumer Electronic Manufacturers Association (CEMA) of the Electronic Industries Association (EIA).
3. The RDS Advisory Group hosted by CEMA.

   4. The RDS Forum of the European Broadcasting Union (EBU).
   5. The FCC's Emergency Alert System Rules and Regulations—47 CFR
      Section 11.

In addition, many organizations, businesses, and radio stations sponsor Internet web sites that provide a wealth of information about RDS. These web sites can be located using many of the existing web browsers.

6

Consumer Receivers

To receive RDS broadcasts, an RDS receiver must be purchased. Over 500 models of RDS receivers are available worldwide today, with many new models being developed specifically for the U.S. market. Automobile manufacturers in the U.S. are beginning to see the value of many of the RDS features that allow greater utility and listening pleasure while driving. Most automobile manufacturers in Europe include RDS in their vehicles as standard equipment. The following types of receivers are available today for purchase by the consumer:

- OEM automotive
- After-market automotive
- Home-stereo tuners and receivers
- Home-theater receivers
- Personal-computer-based receivers
- Portable receivers
- Low-cost receivers, such as clock radios

RDS can be introduced into almost any receiver product that is on the market today. Lets begin our look into consumer products by examining a simple block diagram of a typical RDS receiver, as shown in Figure 6.1.

As can be seen from Figure 6.1, an RDS receiver is just like a non-RDS receiver with the following exceptions:

- *RDS demodulator.* Recovers the data from the 57 kHz FM subcarrier. An asynchronous serial data stream is input to the microcontroller via a clock and data signal.
- *Alphanumeric display.* Since RDS provides a multitude of data to the listener, some type of textual display is required. The minimum display size is eight characters.

33

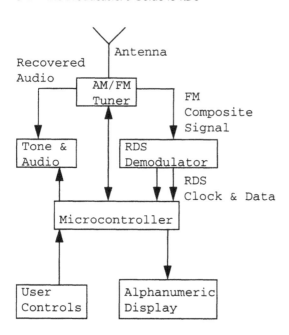

Figure 6-1 Typical RDS Receiver block diagram.

Some of the other receiver sections are also modified to meet additional requirements for RDS. These blocks are:

- *User controls.* Additional push buttons are often added to make selection of RDS features easy to use and differentiate. Some controls may be multiplexed to provide both RDS and non-RDS functions.
- *Microcontroller.* Often times more Read Only Memory (ROM) and Random Access Memory (RAM) are required to implement RDS. The actual amount depends on the number of features included.
- *AM/FM Tuner.* One of the most important features of RDS, the Alternate Frequency (AF) feature requires that the tuner be able to switch from one frequency to another very quickly. Tuners with a very fast Phase Locked Loop (PLL) must be employed to minimize disruption of the audio to the listener.

Consumer RDS Features

A large portion of the RDS features have been designed for listener convenience. These listener features can be incorporated into a receiver in a variety of ways. Not every receiver will have all the features possible, therefore it is important to be knowledgeable when making a purchase so that you get a receiver that has the features you want. The first thing to look for is the

Figure 6–2 The RDS logo. **Figure 6–3** The RDS EON logo.

RDS logo. This logo is shown in Figure 6.2. In the U.S., receivers must be authorized by the National Association of Broadcasters (NAB) in order to carry the RDS trademark. This trademark carries a very minimal set of requirements with it due to the wide range of applications that may or may not be incorporated into a receiver. For this reason, do not choose a receiver by the logo alone. The RDS logo will most often be located on the face of the receiver.

Receivers that are equipped with the Enhanced Other Network (EON) feature will carry a different type of logo, as shown in Figure 6.3. The EON feature enhances the operation of other RDS features, such as traffic announcements and dynamic program type interrupts.

The consumer features most often found in an RDS receiver are:

1. *Program Service (PS) Name.* This is the eight-character name or slogan for the station to which the receiver is tuned. The PS will often replace the frequency information in the receiver display when tuned to an RDS station. When tuned to a non-RDS station, the frequency will be displayed.
2. *Alternate Frequency (AF) Switching.* This feature allows the receiver to automatically tune to the best signal available when the broadcast is carried on multiple frequencies. Many stations carry the same broadcast on several different full-power transmitters or, more frequently, on low-power translators.
3. *Program Identification (PI) Code.* While not exactly a user feature, this code forms the basis for many of the RDS features. The Program Identification Code is the "signature" for your station, allowing the receiver to insure that it is tuned to the proper broadcast at all times. This code is essential to the proper operation of alternate frequency, enhanced other networks, and receiver memory. It is possible to determine a station's call letters in the U.S. based on the program identification code.
4. *Traffic Program (TP)/Announcement (TA).* This feature allows the receiver to respond when a traffic bulletin is given through the main

audio channel. The receiver will switch to the tuner mode when an announcement is given.

5. *Radiotext.* This feature allows reception of either thirty-two- or sixty-four-character messages as transmitted by the broadcaster. This could include, for example, the artist and song title or station promotional information.

6. *Program Type (PTY) Display.* Displays the current program type being broadcast by the tuned station. Up to thirty-one different program types are defined for use by broadcasters. Some examples are News, Sports, and Classical.

7. *Program Type Name (PTYN) Display.* Displays the current PTYN being broadcast on the tuned station. The PTYN is up to eight alphanumeric characters that further describe the PTY. For instance, a station broadcasting a PTY "Classic Rock" may also transmit a PTYN of "70s" to describe the program.

8. *Program Type Search or Scan.* This allows the listener to search across the band for a desired program. The listener would select a program from the list of the available program types and press seek or scan to initiate the search.

9. *Program Type Watch or Interrupt.* This allows the listener to select one or more program types to standby for. As in the case of the traffic announcement feature, the receiver will switch over to the tuner mode when the desired program becomes available.

10. *Emergency Broadcast Alert.* This feature operates just like the program type watch feature, with the exception that the code is predefined and unalterable. Reception of PTY code 31 will cause the receiver to switch over to tuner mode for the reception of an emergency bulletin. Special operation under the Federal Communications Commissions Emergency Alert System is also possible through a special Open Data Application.

11. *Enhanced Other Networks.* This feature adds enhanced capability to other features by allowing the receiver to store and respond to programs on different transmitters or networks, as well as obtain basic tuning data. No tuning will occur to other stations unless your station transmits the EON data. Listeners cannot be stolen away with this feature. EON enhances the operation of the following features:

- Traffic Announcement
- Program Type Search
- Program Type Interrupt
- Program Service Name Display
- Alternate Frequency Tuning
- Program Item Number

12. *Clock Time and Date.* This allows you to broadcast the current time and the Julian date. From this data, the local time, date, day of week, month, and year is known. This can allow the receiver to set its internal clock automatically. This data should only be transmitted if it is accurately maintained (+/ − 15 s).
13. *Music/Speech switch.* This feature allows you to tell the receiver whether the current audio program is music- or speech-oriented. The receiver may use this to modify tone settings or even set the audio to mono to reduce the signal to noise ratio under adverse signal conditions.
14. *Decoder information.* This allows you to indicate to the receiver if the transmitted program type codes are static or dynamic. This provides an important piece of information to receivers incorporating the program type interrupt feature by indicating that the feature may not be available on the tuned station. The decoder information also allows you to indicate that the broadcast is in stereo or mono, compressed or not compressed, or recorded with or without an artificial head.
15. *ID Logic.* ID logic is not an RDS feature, but emulates some of the RDS features through the use of a database built into the receiver. Through this database, the call letters and format of the received station can be determined. Most ID logic receivers allow you to search for stations by format type. The defined formats for ID logic do not entirely match those used in RDS. ID logic is useful for obtaining RDS-like features on AM and non-RDS FM stations.

Selecting an RDS Receiver

There are wide variations in the features and implementation of RDS receivers. From the preceding list you should be able to construct a shopping list of features that are desirable. You can tell a lot about the radio you are looking at by a few simple criteria. The first thing to examine is the display. The display is probably the most expensive component of an RDS capable receiver; look for the following items:

1. *Number of characters in the display.* An RDS receiver requires at least eight alphanumeric characters. The program service name, program type code, and program type name features are all up to eight characters in length. If you are choosing a receiver with the radiotext feature, then a display with more than eight characters is desirable since it will allow the message to be more easily read when displaying long-text messages.

16 Segment   5 x 7

British Flag Character   Dot Matrix Character

Figure 6–4 Typical receiver display types.

2. *Display type.* Either a Liquid Crystal Display (LCD) or Vacuum Fluo-
rescent (VF) display are typically used. LCDs come in a variety of types
and colors. VF displays will typically be blue-green in color against a
black background. Generally speaking, a VF display is more readable
at longer distances and sharper viewing angles than an LCD. Large VF
displays are also generally more expensive than LCDs, often due to the
need for a DC/DC converter to obtain the proper grid voltages.

3. *Type of character generation.* The two basic types of text character
generation are British flag and dot matrix, as shown in Figure 6.4.
These displays are typified by the following characteristics:

- British flag displays create each character through either fourteen or
sixteen segments. Because the character area resembles a British
flag when you turn it on its side and all segments are illuminated, this
display is named as it is. The British flag display is typically harder to
read than a dot matrix display but is less costly to produce.
- Dot matrix displays create each character by illuminating the
proper combination of dots within an array. The most common array size is five columns wide by seven rows deep and is often referred to as a "five by seven" (or 5 × 7) display. This type of display
produces characters that are finer in detail and easier to read, especially at a distance. It is typically more expensive, since many
more segments must be controlled.

The next item to look for is the front panel push-button layout and labeling. The RDS features should be clearly labeled and easy to use. It is prefer-
able that the RDS controls be separate from the other receiver features, but
they may often be multiplexed to reduce the total number of push-buttons.
Look at the RDS push buttons and attempt to determine the receiver operation

before you open the user's manual. Once you are generally satisfied with the layout of the radio, open the owner's manual and verify that all the features that you are looking for are included in the receiver. As a final check, verify that you can operate the features as described in the owner's manual.

Parametric Measurements of RDS Performance

The International Electrotechnical Committee (IEC) has published a set of standards for evaluating the parametric performance of RDS receivers. Since almost every RDS receiver utilizes a VLSI integrated RDS demodulator, the actual performance of the RDS section of the receiver is mostly determined by the sensitivity of the tuner. The RDS sensitivity rating is probably the most important parameter to note. The actual parametric performance of the RDS section may be published in the owner's manual. If so, make comparisons of various competitive sets before deciding on a purchase.

Receiver Operation

Now that we have used this information to make a smart purchase decision, let's look at some typical implementations of RDS. The operations outlined below are only examples of possible implementations. Check out the owners manual for actual operation. For this example we will construct a hypothetical receiver for our test drive, as shown in Figure 6.5.

The first thing we notice, besides the rather large RDS logo, is the addition of six distinctly RDS push buttons. The display contains an eight-character,

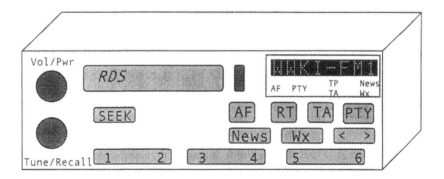

Figure 6–5 Hypothetical RDS receiver.

5 × 7, dot matrix display, as well as several associated icons labeled similarly to the RDS pushbuttons. The operation of the receiver is described in detail below.

Alternate Frequency (AF) Selection

The AF push button turns the Alternate Frequency feature on and off. When turned on, the receiver will automatically switch between transmitters carrying the same broadcast. The receiver will not switch unless:

1. The AF feature is turned on.
2. The RDS transmission carries the proper list of frequencies in the AF list (group 0A).
3. All the transmitters broadcast the same program identification (PI) code.
4. The tuned station's signal quality becomes unacceptable and the signal quality of one of the alternate frequencies is acceptable.

As you can see, the AF feature works to your benefit in allowing FM translators and other full-power transmitters to be linked together. To the listener, your station appears as one large transmitter, since they will not know that the receiver is changing frequency. The listener sees only the PS name of the station, not the frequency. The frequency diversity aspect of RDS is one of its most appealing attributes to the listener, since manual tuning or frequency memorization is eliminated. AF allows FM stations to obtain the coverage area of AM, but with higher quality. When the AF feature is turned on, the AF icon in the display appears so that the listener is aware that the feature is on.

Judging the Quality of the AF Feature

To perform an alternate frequency switch from one transmitter to another, the receiver must evaluate the signal quality of the alternate frequencies. To do this, the tuner must be switched very quickly from one frequency to another, and then back to the original frequency. As shown in Figure 6.6, this alternate frequency check will cause a temporary disruption of the audio. The higher the quality of the receiver, the shorter the length of the mute time. Some receivers may not even have an AF push button if the AF mute times are extremely short. Some receivers are equipped with two tuners to allow silent AF switching. These receivers are more expensive and may suffer some receiver sensitivity degradation, since two tuners are fed with a single antenna.

A different type of audio mute occurs when the receiver actually switches to an alternate frequency. This mute occurs because the receiver must verify that the PI code of the two transmitters match. The receiver must do this to prevent switching to a transmitter belonging to another network. Under some conditions, the receiver will verify the PI code after the audio is

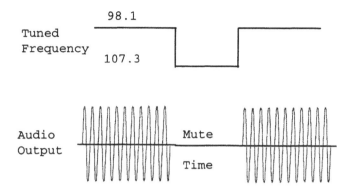

Figure 6–6 Alternate Frequency check.

Status		Receiver Presets						Pool Memory	
Description	Signal Quality	1	2	3	4	5	6	Record 1	Record x
PI Code		8FCC	B102	5431	21FC		7DDA	5C09	BF02
PS Name		WWKI-FM	NPR-1	Q-95	Rock-104		The Edge	WLS-FM	NPR-2
PTY code		6	3	7	20		10	14	21
Preset Frequency	5	97.3	102.7	88.9	103.3	102.3	92.5	102.9	90.7
AF #1	6	**100.5**	103.1	90.1	105.5		87.5	103.9	91.3
AF #2	0	99.5	98.7	91.7	106.7		87.5	104.1	92.1
AF #3	1	103.5	87.5	92.3	87.5		87.5	94.7	89.7
AF #x	3	107.3	87.5	104.5	87.5		87.5	87.5	87.5

Figure 6–7 RDS receiver memory.

unmuted if there is a high degree of confidence that the alternate frequency is in the same network.

Receiver Memory

RDS receivers store much of the data for instant recall during radio operations. Figure 6.7 shows a typical RDS receiver memory.

As can be seen from Figure 6.7, the receiver must keep track of all the received data in order to provide instant tuning to all the networks. When a frequency of 87.5 is entered into the receiver memory, this means that no AF exists. This is referred to as the filler code. In this example, note that preset

1 is the currently selected network. The receiver is currently tuned to AF #1 at 100.5, since it has the best signal quality at present. From looking at this table we can derive the following:

- The PS information is stored in the receiver for instant recall. When the receiver is tuned either manually or through a preset recall, the PS information can be displayed as soon as the PI code is received. This is one reason why the PS information must not be dynamically changed.
- Preset 5 is a non-RDS station; only the preset frequency is stored.
- Preset 6 is an RDS station with only one transmitter.
- The pool memory contains information about RDS stations that are not stored as preset. This provides the same convenience for all RDS stations.
- Preset 2 and pool record "x" contain data about a national network. Special PI codes are assigned for national networks; these PI codes vary only in the second nibble of the PI code.

Program Identification Search

A special case occurs when an RDS preset is recalled and none of the frequencies are acceptable. When this occurs the receiver enters into a special mode called the Program Identification or PI Search; the receiver will seek all stations across the band in an attempt to locate a station with a matching or a variant PI code. For an example, we will recall preset 2, since variant PIs are accepted in the B000 range. When preset 2 is recalled the following events occur:

1. The receiver attempts to tune to any of the AFs stored in preset 2's memory. The receiver tunes to each of the frequencies and measures the signal quality. If no stations are acceptable, the receiver moves to step 2 only if variants are defined for the desired PI code. If they are not, the receiver continues at step 3.
2. Pool memory record "x" is a variant PI code of preset 2. This indicates that the two networks are affiliated in some fashion with each other. Since the two networks do not necessarily carry the same audio program, the receiver cannot AF switch between them; on a preset recall, however, it is acceptable to tune to a variant PI code if the desired network is unavailable. National Public Radio (NPR) is a good example of this. Therefore, the receiver will now attempt to tune to one of the frequencies stored under pool record "x." If none are available, the receiver then moves to step 3.
3. The PI search is now initiated. The receiver display shows "PI Search" or "NPR-1 Search." The receiver tunes to every station on the band and will tune to any station having a PI code of "BX02," where x is any hexadecimal number. If no stations are found, the receiver moves to step 4.

4. Having attempted every possible operation to locate the desired service, the receiver now tunes to the preset's originally stored frequency of 102.7. If there is an RDS station present at 102.7 with a different PI, then the new network will be accepted as the currently tuned network. If the listener recalls the same preset again, the entire operation will repeat.

AF or RDS?

Some receivers may have a push button labeled "RDS" instead of "AF." With this approach the receiver will commonly turn off all RDS features, except the emergency alert feature, when RDS is turned off, making the receiver operate just like a non-RDS receiver.

Radiotext Operation

The listener has access to the broadcast radiotext messages contained in groups 2A or 2B. These messages may be up to sixty-four characters in length. This particular receiver has only eight characters of displayable text at any given time, so the receiver must format the text to properly fit into the display. Two options are available:

1. *Scrolling text.* The text message is scrolled from right to left across the display at a fixed rate. The user reads the message as it scrolls by. This method requires that the user pays close attention to the display for the duration of the message.
2. *Page-formatted text.* The text message will be paged with words formatted into whole words wherever possible. Words over eight characters in length will still be broken from one page to the next. Page formatting is easier to read than scrolling. Messages may be read quickly if the user is given the ability to quickly page through the message.

Figure 6.8 shows an example of the two type of radiotext displays. The Radiotext Message reads, "Now playing on WFBQ: 'The Pass' by Rush."

Traffic Announcements

RDS traffic information may be obtained by pressing the TA push button. Note that two icons appear in the receiver display. These icons provide useful information at a glance:

- TP—Illuminates when the receiver is tuned to a traffic capable station.
- TA—Illuminates when the user has selected the traffic announcement mode.

Radiotext Message: "Now playing on WFBQ: "The Pass" by Rush"

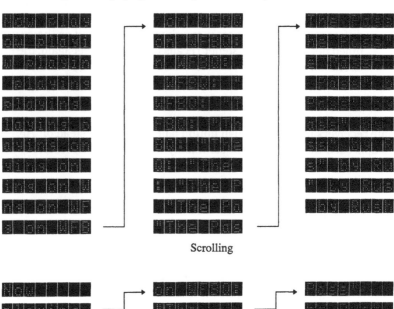

Scrolling

Page Formatting

Figure 6–8 Scrolling versus page formatted Radiotext displays.

If the listener is tuned to a traffic capable station (the TP icon is on), the receiver will toggle on and off the Traffic Announcement feature when the TA push button is activated. If, however, the listener is currently tuned to a non-traffic capable station when TA is activated, many receivers will automatically tune to a traffic station. When TA mode is on, pressing the seek push button will cause the receiver to seek the next traffic capable station. Non-RDS and non-traffic capable stations will be ignored during the seek. This insures that the listener does not inadvertently tune to a station incapable of providing the service that was selected.

Traffic Announcement Volume

Most receivers allow the user to preset the volume level to be used during a traffic announcement. This allows the traffic volume to be set at a consistent level, even though the radio volume may currently be set very loud or soft. Some receivers allow the traffic volume to be set by pressing and holding the TA push button. Other receivers store the last volume level that was used during a traffic announcement to be the default level. In this case, the listener simply adjusts the volume to the desired level during the announcement.

Operation during an Announcement

When a traffic announcement is broadcast, the receiver will stop any current playback operations and switch to the FM tuner mode. The volume level will be set to the user preset level. "Traffic" may be displayed in the receiver display for some time to inform the user why the receiver has changed operations. When the announcement is over, the receiver will return to its prior mode and volume setting. During a traffic announcement, the user may press the TA push button to cancel the current announcement, but without deactivating the announcement setting. If the receiver is equipped with the Enhanced Other Networks (EON) feature, the receiver may be tuned to a different network or station for a traffic announcement. Operation during and after the announcement is the same, with the exception that the receiver is temporarily tuned to another station during the actual traffic announcement. This cannot occur without the broadcaster telling the receiver to switch stations. Listeners cannot be "stolen" with the EON feature.

Program Type Mode

The program type (PTY) mode allows the user to search or scan for their favorite type of program. There are predefined PTYs for all major types of music and talk programs. The use of PTY allows the listener to find the desired program automatically. The listener does not have to tune to each station and listen to the broadcast to determine the type of program. This is often difficult for the listener to determine, especially during commercials. The PTY mode is activated as follows:

1. The user turns on the PTY mode by pressing the PTY push button.
2. The user scrolls through the choice of available PTYs by pressing the up/down arrows ("↑↓") until the desired PTY is displayed.
3. The user then presses the seek or scan push button to begin the search. The receiver will then stop on the first station broadcasting the desired PTY.

Some receivers allow the PTY mode to be "latched." Latching the PTY mode allows the receiver tuning to be totally altered based on the current selected PTY. For example, the listener turns on the PTY mode and tunes to a Rock station using the previously described procedure. The listener then decides to tune to another Rock station. The user simply presses the seek push button to find the next Rock station. The user can instantly see that the receiver is in the PTY tuning mode by the status of the PTY icon in the display. Some receivers may allow the preset push buttons to be defined as PTYs when the PTY mode is on. This then allows the listener to store and instantly recall his favorite PTYs to be used in search tuning.

Program Type Watch or Interrupt Mode

To take maximum advantage of PTYs, the broadcaster should dynamically change the PTY with the audio program. Some receivers are equipped with a special mode that allows the listener to select one or more PTYs to "watch" for. The operation of the receiver is very similar to that during a traffic announcement. With our hypothetical receiver, we are allowed two choices for PTY interrupts, News and Weather. Icons for News and Weather will be illuminated in the display when the user has selected one or both to watch for. The receiver will watch for the received PTY to change to the one selected and will switch over to the tuner mode for the duration of reception of the desired PTY. Analogous to the EON operation of traffic announcements, the receiver can switch over to another station when the desired PTY becomes available. The main difference between the traffic and PTY features is that the listener has no idea whether or not the desired PTY will ever be broadcast. There is no bit analogous to the TP code for PTYs. The only feature that can be of help is the dynamic PTY indicator that is contained within the decoder information bits. When set, the dynamic PTY indicator informs the receiver that the tuned station dynamically switches its PTY code. Thus, there is at least a chance of receiving the desired PTY. Our receiver can use this information so that when the user presses the News push button to activate the PTY interrupt feature, the receiver could display "Not available."

Program Type Name (PTYN) Display

A special feature allows the broadcaster to send up to eight alphanumeric characters that describe the program content in finer detail than the PTY allows. For instance, when the user first tunes to an RDS station, the PTY is displayed:

The receiver then decodes and displays the PTYN:

Then the PS Name is displayed:

While the listener cannot search for PTYN, it does allow them to quickly get further information about the broadcast. It also allows the broadcaster to clearly identify itself from the two other classic rock stations in the market. This and other information can often be redisplayed by pressing the recall push button on the receiver.

Emergency Warning Feature

A special PTY code is reserved for emergency announcements. Many receivers that may not have the PTY watch mode will still have the emergency alert feature. When the receiver receives a PTY code 31, the tuner mode will be activated. It is possible that our receiver may be configured to turn itself on when a PTY 31 is received. A special Open Data Application (ODA) has been defined to augment the emergency alert feature. This special ODA allows consumer equipment to automatically detect stations that provide the emergency alert feature. Additionally, receivers may use a sleep/wake cycle to conserve batteries when monitoring this ODA. Operation with our hypothetical receiver could be as follows:

1. The listener is driving on a long trip and decides to turn the receiver off.
2. The receiver automatically seeks across the band until the desired ODA is detected. The receiver then enters a watch mode, looking for the PTY code to be set to 31.
3. If a PTY code equal to 31 is received, the receiver will turn on and set the audio to the traffic announcement volume. The receiver display shows "ALERT!" and immediately notifies the listener, who has glanced over to figure out why the receiver has switched on.
4. The listener receives an audible message that a flash flood warning has been issued for several counties until later that evening. The listener then decides to alter his course away from the affected areas.
5. At the conclusion of the audible warning, the receiver detects that the PTY has been set to something other than 31 and turns itself back off.
6. A similar bulletin has been entered into the radiotext, allowing those who tune into the broadcast after the audible announcement to receive a similar warning.

Unlike the PTY defined for weather, the alert PTY is defined only for emergency announcements, rather than the daily forecast.

Home Receivers

RDS also brings benefit to non-mobile receivers. Many home tuners, home receivers, and home theater receivers now incorporate RDS. Even the Alternate Frequency feature can be put to good benefit in a home receiver by incorporation of the FM Memory feature. The FM Memory feature presents the same program appearing on multiple frequencies as the same program by only tuning to the strongest station when that particular program is selected. The PTY interrupt mode also allows for the great benefit of another RDS feature that normally would be of limited benefit in a non-mobile application. Figure 6.9 shows a typical RDS home receiver.

Another novel application for RDS is with home computers. An RDS receiver can be fitted within a personal computer. This is especially useful for data transmission as well as audio entertainment. RDS offers a low-cost, convenient method for data transmission without the requirement of a dedicated

Figure 6-9   A RDS home receiver. (Photo courtesy of Denon Electronics.)

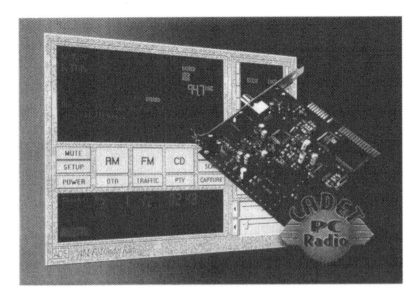

Figure 6-10 A PC-based RDS receiver. (Photo courtesy of Advanced Digital Systems.)

phone line. A PC based RDS receiver is shown in Figure 6.10. This receiver offers AM/FM Stereo RDS reception and also includes the ID Logic feature. A special data service including news, paging, and E-mail notification is also offered through the use of MBS.

Program Identification Codes and Extended Country Codes

7

The Program Identification or PI code is the backbone of all RDS features. While the listener never sees the PI code, it is the single most important feature used by the receiver to carry out many of the consumer features. The PI code is the signature of the broadcast network. This signature is represented by a binary-coded four-digit hexadecimal code. Consumer receivers identify that the broadcast on two different transmitters is the same by identifying that the PI codes are exactly the same. The PI code is essential for carrying out the following functions:

1. Alternate frequency switching
2. Enhanced other networks features, including:
   a. Traffic announcements
   b. Program Type interrupts
   c. Program Item Number switching
   d. Receiver memory updates of associated networks
3. RDS preset recalls, including PI search
4. RDS receiver FM memory function

Since the PI code is so essential to the proper performance of consumer receivers, it is located in the first block of every RDS data group and is additionally located in block three of every B- type data group. Since the PI code is located within every group in the same location, receivers can decode the PI code without any reference to the group type, as long as the group offset is known.

Determining the PI Code for Your Station

In all areas licensed by the U.S. Federal Communications Commission, the PI code is calculated from the station's call letters. The *RBDS Standard* includes the conversion formulas to calculate your station's PI code. Many RDS encoders have the conversion formulas built into the computer interface software, so that you only have to enter your station's call sign. I will not duplicate the conversion formulas, since they are located in Annex D of the *RBDS Standard*, but do pay careful attention to the exceptions to the calculated PI codes that follow.

Exceptions to the Calculated PI Codes

The formula conversion from call letters to PI codes applies in almost all cases, with some exceptions. It is critical that the following exceptions be adhered to in order to ensure proper operation of consumer receivers:

1. *PI codes that map with a "0" in the second nibble.* PI codes such as 6012 must be remapped to the A _ _ _ range according to the following rule:
   a. P1 0 P3 P4 changes to A P1 P3 P4. Our example would change from 6012 to A612.
   b. Rationale: Some European RDS receivers will not AF switch if a "0" is in the second nibble of the PI code.
2. *PI codes that map with "00" in the last two nibbles.* PI codes such as 7100 must be remapped to the A F _ _ range according to the following rule:
   a. P1 P2 0 0 changes to AF P1 P2. Our example would change from 7100 to AF71.
   b. Rationale: Some European receivers will enter a test mode if the last two nibbles of the PI equal "00."
3. *Three-letter call signs.* The PI code for all stations having three-letter call signs is located within a table in Annex D. Do not use the conversion formula.
4. *Two stations carrying identical programming but having different call signs.* This is covered in detail in this chapter.
5. *Nationally or regionally linked radio stations with different call letters.* This topic is also covered in detail later in this chapter.

Choosing a PI Code for Stations Carrying Identical Programming

In the case where several full-power stations carry identical programming, it is essential that the same PI code be used on all stations. It does not matter which PI code is used, as long as they are all the same. The example shown in Figure 7.1 depicts such a network. In this network the calculated PI codes for the stations are:

- WABC - 54C4
- WDCS - 5CDA
- WZSQ - 9890

The PI code for WABC was randomly chosen from the three available PI codes to be used throughout the transmitter network. If each transmitter operated with a different PI code, then consumer receivers would not switch from one transmitter to the next as the listener drove between the various coverage areas. Consumer receivers use the PI code to verify that they are tuning to the same audio program. Identical PI codes tell the receiver that the audio program is the same. Different PI codes inform the receiver that the audio program is different; the receiver will not switch to a different PI code regardless of the network frequencies transmitted in the alternate frequency (AF) list. Hence, proper

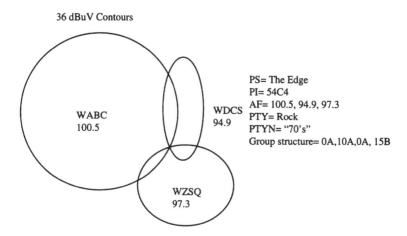

Figure 7-1 PI Code usage for multiple transmitters carrying identical programming.

usage of the PI code is essential to the AF feature of RDS. The PI code is only used internally by the receiver; the consumer never sees the broadcast PI code.

Choosing a PI Code for Nationally or Regionally Linked Radio Stations

A special set of PI codes are reserved for use in the United States, Canada, and Mexico for programs that are nationally or regionally linked. Some examples are National Public Radio (NPR) and the Canadian Broadcasting Corporation (CBC). These special PI codes are interpreted by consumer receivers in a way that is different from all other PI codes. The second digit or nibble of these PI codes contain information about the coverage area of the audio broadcast. Receivers use this information when determining which transmitters to AF switch to in the network. In North America, these PI codes are defined in the range of B_01-B_FF, and D_01-E_FF, as shown in Figure 7.2. These regionally linked PI codes are shared among the U.S., Canada, and Mexico. The administrators of the National Radio Systems Committee, either the NAB or EIA, are responsible for allocating these PI codes.

How Regional PI Codes Are Utilized

The second nibble of the regional PI codes tell the receiver how the tuned station is linked to a larger network. Based on this, the receiver determines which stations the receiver should tune to during an AF switch. The reason that these special PI codes are used is that many large networks have stations that will break away from the main program at different times of the day to provide local programs. Some examples are news, weather, special sporting events, or even locally oriented music programs. While these stations are still part of the larger network, the audio program can differ during these periods of local broadcasting. It may be undesirable to the consumer to have the receiver switch from one station to the next when the audio content is different. It is desirable, however, that the receiver tune to some station in the network when the receiver preset is recalled, even if the program content is local. It may be the only station within listening range at that time.

Figure 7.3 shows the definition of the second nibble of regional PI codes. This regional nibble, often referred to as the regional variant, informs the receiver about the coverage area of the tuned station.

NATIONALLY/REGIONALLY LINKED RADIO STATIONS	HEX CODE = FOUR DIGIT PI CODE
NPR	B_01
CBC - English	B_02
CBC - French	B_03
?	B_04
:	:
?	B_FF
?	D_01
?	D_02
:	:
?	D_FF
?	E_01
?	E_02
:	:
?	E_FF

Figure 7–2 Network Program Identification Codes.

Area coverage code	L	I	N	S	R1	R2	R3	R4	R5	R6	R7	R8	R9	R10	R11	R12
HEX	0	1	2	3	4	5	6	7	8	9	A	B	C	D	E	F

Bits b_{11} to b_8:Second nibble or digit of the PI code

Figure 7-3 Coverage area codes for nationally allocated program identification (PI) codes.

I: (International)	The same program is also transmitted in other countries.
N: (National)	The same program is transmitted throughout the country.
S: (Supra-regional)	The same program is transmitted throughout a large part of the country.
R1 . . . R12: (Regional)	The program is available only in one location or region over one or more frequencies, and there exists no definition of its frontiers.
L: (Local)	Local program transmitted via a single transmitter only during the whole transmitting time.

How Consumer Receivers Utilize Regional PI Codes

Most consumer receivers allow the listener to select whether changes in the coverage area code will be accepted or not during an Alternate Frequency switch. The selections are:

- *Accept variants.* The receiver will accept stations carrying a different coverage area code when performing an AF switch or preset recall.
- *Reject Variants.* The receiver will only accept stations carrying an identical PI code when performing an AF switch or preset recall.

Receivers that do not allow the listener to select the variant setting must be set to reject variants. A case can be made, however, that receivers should accept variants on a preset recall or PI search. When either of these operations are carried out, the listener has selected an action to take place and the receiver audio is muted during this time. If no exact PI matches can be located, then it is acceptable to tune to a regional variant rather than nothing. This is covered in detail in the chapter 13.

A Network Study: The Canadian Broadcasting Corporation

The Canadian Broadcasting Corporation, or CBC, contains two different national networks, an English and a French network. Each network simulcasts at different times of the day, but various stations will break from the network to provide local programming. In this example, we will utilize the regional PI codes that have been assigned to these networks:

1. CBC—English: PI B_02, Assume PS "CBC - E"
   a. All transmitters will use PI code B202 when linked to the network. The second nibble set to "2" means that this is a national network.
   b. When a station or a group of transmitters breaks from the network, the PI code must be set to a regional variant. The second nibble will be set to a value of 4-F (i.e., B402), depicting regional coverage.
2. CBC—French: PI B_03, Assume PS "CBC - F"
   a. All transmitters will use PI code B203 when linked to the network. The second nibble set to "2" means that this is a national network.
   b. When a station or a group of transmitters breaks from the network, the PI code must be set to a regional variant. The second nibble will be set to a value of 4-F (i.e., B403), depicting regional coverage.

For our example, the listener is tuned to CBC-E and is traveling from point A to point B, as depicted in Figures 7.4 and 7.5. A listener has set the preset feature of the receiver to this network. The receiver has been set to reject variants.

In Figure 7.4, all the transmitters are linked to the network, for the PI codes have been identified as equal to B202. As the listener travels, the receiver will tune in the following sequence: 95.5, 106.9, 101.9, and 92.3. The listener has enjoyed the program without interruption throughout the entire journey.

In Figure 7.5, we notice that the network chain is broken along the route from A to B, because two transmitters are broadcasting a regional variant of B602. From this we can deduce that:

1. Transmitters 95.5, 94.7, and 92.3 are all broadcasting the national program.
2. Transmitters 106.9 and 101.9 are broadcasting a regional program.
   a. Both of these transmitters are broadcasting the same program, since they are broadcasting the same regional code, or "6."

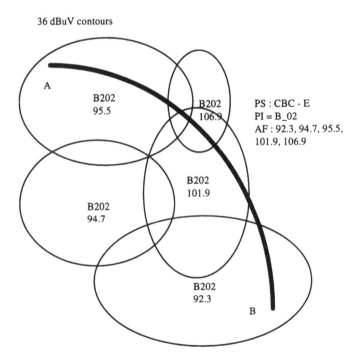

Figure 7–4 Nationally linked program.

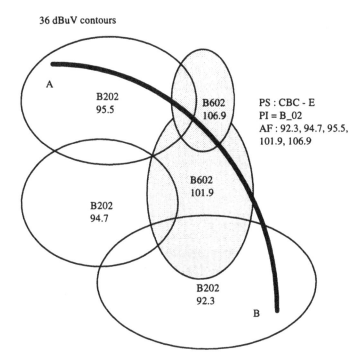

36 dBuV contours

A

B202
95.5

B602
106.9

PS : CBC - E
PI = B_02
AF : 92.3, 94.7, 95.5,
101.9, 106.9

B602
101.9

B202
94.7

B202
92.3

B

Figure 7–5   Nationally linked program with regional variants.

As the listener travels from A to B, the receiver will begin by being tuned to 95.5. As the listener travels out of range of this transmitter, the receiver will attempt to tune to 106.9. The receiver will not tune to 106.9, because the transmitter is broadcasting a regional variant PI code and the listener has selected "reject variants" on his receiver. Likewise, the receiver will not tune to 101.9. The receiver will remain on 95.5 until the point that the listener travels within range of 92.3 or the listener recalls the network preset. If the preset is recalled, the receiver will accept the variant PI code and tune to 106.9. The receiver will then tune to 101.9; but it will not tune to 92.3, since it is broadcasting a PI code that differs from the currently tuned network. Some receivers may AF switch to 92.3 in this instance, because the transmitter is broadcasting a coverage area code that represents a nonregional network. Hence, some receivers may accept an AF switch from a regional variant (4-F) to a nonregional variant (1-3).

If, in this example, the listener has selected "accept variants," the receiver would AF switch just as it did in the first example in which all the transmitters were broadcasting the national coverage area code. The audio content will change as the receiver switches from 95.5 to 106.9, and again when switching from 101.9 to 92.3. The choice is the listener's as to whether "reject variants" or "accept variants" setting is most desirable.

The implementation of this feature is somewhat confusing, due to all the possible scenarios that can occur. In Germany, for instance, the PI codes cannot be dynamically changed from the national code to the regional variant code, even though these regional stations may be linked to the national program for most of the day. This presents a special problem for receiver manufactures who must design a product that is easy to use, but works optimally in all circumstances. It is highly recommended that the regional PI codes be dynamically switched to represent the current audio program at all times.

Multiple Regional Programs

In areas where multiple regional programs exist, it is important that different regional coverage codes be used where transmitters have overlapping coverage areas. The same regional PI code can only be utilized where the same regional program is being carried. Since regional coverage codes are assigned from only 4 through F, advance planning will be required to prevent problems.

PI Codes for Canada and Mexico

For non-networked programs in Canada and Mexico, special PI codes have been assigned, as shown in Figures 7.6 and 7.7. These PI codes function just as the PI codes below B000 do in the United States. AF switching can occur, but regional variant switching is not allowed. Contact the National Radio Systems Committee for PI assignments in these countries.

CANADA RADIO STATIONS	HEX CODE = FOUR DIGIT PI CODE
?	C000
?	C001
:	:
?	CFFF

Figure 7–6 Program Identification Codes for Canada.

MEXICO RADIO STATIONS	HEX CODE = FOUR DIGIT PI CODE
?	F000
?	F001
:	:
?	FFFF

Figure 7-7   Program Identification codes for Mexico.

Extended Country Codes

Outside North America, all PI codes are structured as shown in Figure 7.8. This is the same method by which regional PI codes are structured in North America (see Figure 7.2). What can be seen here is that there are many more countries than there are country codes. What is needed is a way to identify the country of the broadcast. If the exact country of broadcast can be identified by the receiver, then a receiver can be designed that can self-configure itself to tailor its setup for that specific country. This allows for the manufacture of one receiver that can be utilized globally. Such receivers have been developed and are popular in automobiles.

A global receiver is advantageous since it allows for a lower price to the consumer. Global receivers can be offered at a lower price because they can be manufactured for a lower cost. Lower costs can be realized in the following ways:

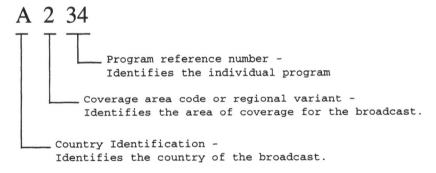

Figure 7-8   Program identification code structure outside North America.

- Inventory is reduced, since only one stock or part number per model is required.
- Mass production techniques can be optimized with fewer line changes.
- Greater yields and fewer defects will be easily realized.
- Fewer hardware and software design errors will exist, causing fewer reworks and noncost-effective fixes.

The use of the PI code along with the extended country code or ECC can allow the receiver to self-configure some of the following parameters:

- AM and FM band edges and tuning steps.
- FM stereo de-emphasis.
- Program Type (PTY) code definitions.
- PI code differences regarding the use of regional variants.
- MBS and MMBS compatibility.
- Open Data Applications, such as the Emergency Alert System or EAS.
- Default language displays.

These things are possible if each broadcaster transmits the extended country code within the RDS group structure. Each country has been assigned a specific ECC for use. These country codes are located in Annexes D and N of the *Standard*. The ECC codes for use in North America are shown in Figure 7.9.

Location of Extended Country Codes

The extended country code information is contained within group 1A, as depicted in Figure 7.10. To transmit this group type, the group structure will have to include group 1A. The ECC information should be transmitted at least once every minute.

COUNTRY/AREA	ISO CODE	SYMBOL FOR PI	ECC
Canada	CA	B, C, D, E	A1
Mexico	MX	B, D, E, F	A5
Puerto Rico	PR	1..9, A, B, D, E	A0
United States of America	US	1..9, A, B, D, E	A0
Virgin Islands [USA]	VI	1..9, A, B, D, E	A0

Figure 7–9 Extended country codes for North America.

Usage	Bit allocation in Block 3															
Group type 1A	b_{15}	b_{14}	b_{13}	b_{12}	b_{11}	b_{10}	b_9	b_8	b_7	b_6	b_5	b_4	b_3	b_2	b_1	b_0
Variant code 0	LA	0	0	0	OPC								Extended Country Code			

Figure 7-10 Structure of extended country code.

8

Program Service Name

The Program Service (PS) Name is your station's new identity with RDS. With the PS feature you are now freed from the chains of frequency and given the latitude to better describe yourself. Up to eight alphanumeric characters may be defined as your station name. Some examples are shown in Figure 8.1. While most stations tend to build their moniker or slogan around their allocated frequency, you now have the latitude to do otherwise. Note that some PS names still reflect the frequency, while others simply state the call sign. Probably the most creative way to market your station are the examples that do not contain references to the frequency or call sign of the station.

Marketing Your Station through the PS Feature

When a listener tunes to your station, the first thing they will see is your PS name. You spend thousands of dollars trying to identify yourself to your listeners each year. You do this in the form of billboards, buttons, banners, balloons, bumper stickers, coffee mugs, T-shirts, giant mobile boom boxes, station vans, and so on. While the station frequency is the dominant idea you want the listener to remember, you can go to extremes trying to find a catchy way to engrave that idea into the listeners head. Catch-phrases or slogans are most often used, such as "Z-93" (92.7) or "Rock-104" (103.9). In the unlucky event the station winds up with a frequency allocation at 500 kHz, you can abandon all references to frequency and come up with "WHEELS" (100.5), for example. Regardless of the slogan, it is now the job of the promotions director to make the listening public aware of his or her station.

* Q-95 *	Wheels
The Zoo	Z-100
ABC-1	WMRI-FM
Shine 99	99.5 ZPL

Figure 8-1 Typical Program Service (PS) Names.

How convenient it would be to have that fancy slogan automatically displayed when any listener tuned to the frequency. Well, with the RDS PS feature that is exactly what will happen. With a one-time investment for the RDS encoder, you will reap constant benefits by this simple yet powerful feature.

How the PS Interacts with Other RDS Features

The importance of the PS Name lies in its use when combined with some other important RDS features, such as the Program Identification (PI) code and the Alternate Frequency (AF) feature. As discussed in Chapter 7, the PI code is the station's signature; it is a unique code that no other broadcast station has. This code is used internally by RDS receivers that actually identify your station by the PI code rather than the frequency. The Alternate Frequency feature allows you to simulcast on multiple transmitters on different frequencies. These transmitters may be full-power stations or translators. The power of the PS lies with these features. Most RDS receivers store RDS stations differently than non-RDS stations. Figure 8.2 shows the memory of a typical RDS receiver. Note that all the RDS stations are actually stored as a PI code; then the PS and the frequencies the service is available on are stored. The non-RDS station is simply stored as a frequency.

When the listener tunes to an RDS station, the stored PS will be displayed as soon as the PI code is received. The PI code will be received much quicker than the PS due to the fact that it takes fewer bits to transmit and is repeated more frequently. Also note from this example that most of the RDS stations were available on multiple frequencies. The receiver will tune to the best available station in the list each time the radio preset is recalled and as the listener travels from one transmit coverage area to another. The one thing that remains constant is the PS name of the station. This makes sense from the listener's point of view, because the same broadcast program is being received regardless of the tuned frequency. Frequency has become unimportant to the listener; to your advantage. In the case of WMRI-FM, shown in Figure 8.2, the listener now perceives four separate transmitters as one single, very powerful station! That is one sure way to improve listenership.

Preset Number	1	2	3	4	5
PI Code	8FCC	761A	B101	520A	-
PS Name	Rock-104	WMRI-FM	NPR	Wheels	-
Frequency	103.9	106.9	92.1	100.1	94.7
AF's	102.9	95.3	90.7		
		102.7	91.3		
		96.9	93.1		
			89.7		
			92.3		

Figure 8-2   Typical RDS Receiver Memory.

Rules for Implementing the PS Feature

The PS Feature is intended to denote the Program Service Name. This name is not dynamic in nature but can be changed when necessary. There has been some use of the PS feature in a dynamic mode, often in an attempt to emulate the dynamic text feature, Radiotext (RT). The RDS specification strictly forbids the dynamic use of the PS feature. The difference between the implementation of the PS and RT features is that the PS is always displayed on the receiver, while the RT message will be displayed only when the listener deems it safe to do so. Imagine, if you will, driving down the road with a constant flashing text message across your receiver. Dynamic PS is a distraction and a safety hazard to the listener. In Europe, where RDS has been in use for over ten years, there have been a growing number of complaints where dynamic PS is used. If the PS is constantly changed, it is possible that some receivers will display a mixture of old and new PS information on the same display. In some cases, no PS will be displayed at all.

As mentioned earlier, consumer receivers often store the PS in memory for instant recall upon receipt of the program identification (PI) code so that the listener has quick access to the station name when tuning to the station. If the PS is used dynamically, whatever was last received will be stored in memory and displayed when the user returns to the station or presses the receiver preset recall button. This often results in a meaningless display being shown on the receiver, totally defeating the purpose of the feature. Dynamic text information should be transmitted using the RT feature in which the listener can control the display of the information for the safest time. Receiver manufacturers are including the RT feature in most receivers, so there is no lost benefit to the station.

Choosing a PS

The choice of the PS is strictly a personal one. Choose a PS that best matches the promotion identification that you currently give to your listeners. It is best to avoid the use of frequency within the PS, since with RDS frequency is almost irrelevant to the listener. If you implement the Alternate Frequency (AF) feature, the receiver will automatically switch frequencies between transmitters in your network to obtain the best reception. Even if you do not implement this feature, frequency changes are often necessary to obtain coverage increases. An added bonus is that RDS receivers will automatically locate your station if you should change frequencies by use of the Program Identification (PI) search feature. This makes the use of frequency references in the PS even less desirable.

Transmitting the station call letters in the display may seem safe, but it is not terribly exciting. If your call letters are your main identification to your listeners, try creating something colorful and catchy for the PS. For example, a station using "WWWO-FM" might use "*3_W_O*" for the PS instead. Those of you with vanity license plates on your car can surely aid in this process.

Character Limitations

The ability to represent characters on a receiver display is dependent upon two basic criteria:

1. The type of display being either a 14- or 16-segment British flag, or a 5 × 7 dot matrix. The dot matrix display has the ability to represent many more characters than the British flag, but is more expensive.
2. The amount of available character ROM. The dot matrix display may have the ability to form many characters, but the display driver IC must have enough memory to store all the representations in ROM. The amount of ROM available is limited due to cost restraints.

If a received character is not displayable by the receiver, it will be represented by a space in that character position. In the case of the previous example—"*3_W_O*"—it would be discovered that a British flag type display would present that PS as "_3_W_O_" on the receiver. As can be seen from this example, either version of the display would be deemed acceptable simply due to the placement of the special characters "*" outside the meaningful portion of the text. See Chapter 11 for more details on text presentation on consumer receivers.

9

Program Type and
Program Type Name

The Program Type (PTY) and Program Type Name (PTYN) features have been designed to allow you to advertise the single most important aspect of your broadcast—your program. The format that you have chosen for your station is the lifeblood of your investment, because your ratings are greatly determined by the format you broadcast. You have also built a listenership by offering a variety of services for your listener. News, weather, sports, and traffic information are an important part of your overall programming as well. Often it is the quality of these services that will keep people coming back. RDS offers you the ability to advertise your program to your listeners with the PTY and PTYN feature. The definitions for the program types is one area in which the European *RDS Standard* and the North American *RBDS Standard* differ.

How to Use the PTY Feature

The PTY feature is utilized by selecting a predefined fixed code that best describes the current program material. There are twenty-six codes available to choose from for use in North America, as shown in Table 9.1. Outside of North America, the program types are defined as shown in Table 9.2. A full description for each PTY follows each table. Codes have been defined to represent music, talk, and service-related programs. Only one code can be chosen at a time. Depending on the degree of automation at your station, you may use the PTY either statically or dynamically. If you are using the PTY in

Number	Program type	Display
0	No program type or undefined	_****_
1	News	NEWS
2	Information	INFORM
3	Sports	SPORTS
4	Talk	TALK
5	Rock	ROCK
6	Classic Rock	CLS_ROCK
7	Adult Hits	ADLT_HIT
8	Soft Rock	SOFT_RCK
9	Top 40	TOP_40
10	Country	COUNTRY
11	Oldies	OLDIES
12	Soft	SOFT
13	Nostalgia	NOSTALGA
14	Jazz	JAZZ
15	Classical	CLASSICL
16	Rhythm and Blues	R_&_B
17	Soft Rhythm and Blues	SOFT_R&B
18	Language	LANGUAGE
19	Religious Music	REL_MUSIC
20	Religious Talk	REL_TALK
21	Personality	PERSNLTY
22	Public	PUBLIC
23	College	COLLEGE
24-28	Unassigned	
29	Weather	WEATHER
30	Emergency Test	TEST
31	Emergency	ALERT!

Table 9-1
North American Program Types.

a static mode, it should be set to the general format of the station. If you have the ability to dynamically vary the PTY, the code should be set for the current program being broadcast.

Special PTY Codes

Several PTY codes are utilized for special purposes. These codes are implemented differently in consumer receivers than the normal PTY codes.

Program Type 0. Program type "0" is defined as "No program type, or undefined." Use this code when you do not wish to define your program at all. This may be appropriate when you offer a wide variety of programs and do not have the means to dynamically change the transmitted PTY.

Program Types 24-28. In North America, these program types have not been assigned. Do not transmit these codes.

Number	Program type	8-character display
0	No program type or undefined	None
1	News	News
2	Current Affairs	Affairs
3	Information	Info
4	Sport	Sport
5	Education	Educate
6	Drama	Drama
7	Culture	Culture
8	Science	Science
9	Varied	Varied
10	Pop Music	Pop M
11	Rock Music	Rock M
12	M.O.R. Music	M.O.R. M
13	Light classical	Light M
14	Serious classical	Classics
15	Other Music	Other M
16	Weather	Weather
17	Finance	Finance
18	Children's programs	Children
19	Social Affairs	Social A
20	Religion	Religion
21	Phone-In	Phone In
22	Travel	Travel
23	Leisure	Leisure
24	Jazz Music	Jazz
25	Country Music	Country
26	National Music	National M
27	Oldies Music	Oldies
28	Folk Music	Folk M
29	Documentary	Document
30	Alarm Test	TEST
31	Alarm	Alarm !

Table 9-2
European Program Types.

Program Type 30. This code should be transmitted only when performing an emergency warning system test. It cannot be used by the receiver in the PTY search mode.

Program Type 31. This code should be transmitted only when performing an actual emergency alert. The PTY should be set to 31 during the actual reading of the alert information. Most consumer receivers will respond to the reception of this code by interrupting cassette or CD playback and switching to tuner mode for reception of the audible alert. It cannot be used by the receiver in the PTY search mode. In the U.S., use of this code has been further defined for use within the Emergency Alert System. See the section on "RDS and the Emergency Alert System (EAS)" in Chapter 15, or Annex R of the *RBDS Standard*, for a detailed explanation.

Newly Defined PTY Codes

With the recent updates to the European *RDS Standard* and the North American *RBDS Standard*, previously undefined program types have now been defined. In North America, PTY codes 23 ("College") and 28 ("Weather") have been added. Outside of North America, program types 16-30 have been added. These codes have been chosen to insure compatibility with the Eureka 147 Digital Audio Broadcasting standard. Older consumer receivers may not be able to decode or use the newly defined program types.

How an RDS Receiver Uses the PTY Feature

The PTY information is transmitted as a binary code and is contained in every RDS data group. The consumer receiver quickly determines the text equivalent of the code by utilizing an internal look-up table. The receiver will display the eight-character representation for the program, as shown in Tables 9.1 and 9.2.

There are several different methods by which a consumer receiver can use the PTY information. Many RDS receivers have the ability to search selectively or scan for stations with a particular PTY. This is referred to as a PTY seek. Another use of the PTY is to "standby" until the desired program of the selected station becomes available.

Program Type Seek or Search Mode

The PTY seek mode allows the listener to find a desired program quickly and automatically. The user selects the desired program from the list of available codes and then invokes the receiver-scan function. The receiver then performs the following functions:

1. The receiver seeks up to the next station that exceeds the minimum signal strength threshold.
2. The receiver attempts to synchronize to the RDS data.
3. If the receiver cannot obtain synchronization within a predetermined amount of time, or obtains a nonmatching PTY code, then it will seek up to the next station.
4. If the receiver obtains the desired PTY code, then the seek will be ended. Normally the PTY will then be displayed for a few seconds.
5. If the receiver seeks through the entire band and finds no matching PTY, then the PTY seek will be ended.

The PTY seek feature allows the user to tune only to stations with the desired program. Normally, a listener would have to listen to all stations, especially in an unfamiliar area, to find the desired program. Often times the listener will tune past a station with the desired program if a commercial is playing at that time. With RDS, the receiver will stop on the station, even during a commercial, if the desired PTY is located. Additionally, the PTY is often displayed on a consumer receiver whenever a station is first tuned, allowing yet another way to advertise your program material.

Program Type Watch or Interrupt Mode

Another use of the PTY code in consumer receivers is similar to the operation of the Traffic Announcement (TA) feature. The PTY Interrupt or PTY Watch mode allows the user to select one or several PTYs that will interrupt cassette or CD playback when the selected PTY becomes available. The operation is as follows:

1. The listener tunes his receiver to the desired station.
2. The user selects one or more program types that the receiver will standby for.
3. The listener can then select a cassette or CD as the source. The listener can also turn the volume down slightly or entirely mute the audio.
4. When the tuned station changes from the current PTY to the listener's desired PTY, then the receiver will:
   a. Stop the cassette or CD playback if in operation.
   b. Change the volume to a user preselected level.
   c. Switch over to the tuner mode. Generally the Program Service (PS) name and PTY are then displayed.
5. When the PTY changes to an unselected code, then the volume level will return to the prior level and the receiver will switch back to the cassette or CD playback, if this was the prior mode.
6. If desired, the user can end a PTY interrupt in progress and standby for the next one.

Program Type 31, Alert Mode

Most consumer receivers have a special feature that responds to the reception of a PTY code 31. This feature generally cannot be over-ridden by the user. PTY code 31 is transmitted whenever an emergency bulletin is being given by the tuned broadcast station. Operation of the receiver is similar to that of the Program Type Interrupt mode, in that cassette or CD operation will be suspended and the audio will be switched over to the FM tuner for the duration of the reception of PTY code 31. Incorporation of this feature is independent of the PTY Interrupt mode.

The Dynamic PTY Indicator

A newly defined use of the decoder information feature is the addition of a dynamic program type indicator. This bit should be set high when the program type is dynamically switched, or when PTY information is updated via the Enhanced Other Networks feature. Receivers that incorporate the PTY interrupt mode can use this information to indicate to the user that the feature is not available on the tuned network. If the program type is statically set, then the dynamic PTY bit should be set to 0.

Limitations of the PTY Feature

While the Program Type feature is very powerful, some important limitations should be noted. One of the most important is that there is no way to indicate to the listener what programs will be broadcast by the tuned station. So while a listener may select a PTY Interrupt of "News," for instance, there is no assurance that the tuned network will ever broadcast that program type. Hence, the user will wait indefinitely for the desired program. The dynamic PTY indicator solves part of the problem by at least indicating that PTYs are dynamically switched. The user will learn by trial and error which stations offer the programs that he or she desires. A better way to solve this deficiency is for you to advertise the features and programs you offer through RDS. This will not only benefit the listener, but will help promote the use of RDS and increase the listenership of your station.

The Program Type Name (PTYN) Feature

The Program Type (PTY) codes have been carefully chosen to represent programs that do not change over time so they will not become outdated. In reality, many different variations of Rock, Pop, and Top-40 come and go in popularity. How, then, are these variations accounted for when only fixed PTY codes are defined for use? The Program Type Name (PTYN) feature allows you to further describe the PTY with up to eight alphanumeric characters. Some examples of this are shown in Table 9.3. As can be seen in the examples, the PTYN feature allows you to fully define the program in order to differentiate your broadcast from all others. Most major markets will have at least several stations with the same general format, but each station

PTY	PTYN 8 Character	Description
Rock	Altrntiv	An Alternative Rock format
Classical	Opera	A classical format station currently airing an opera
Personality	P Harvey	A program featuring radio personality Paul Harvey
Sports	Colts	A live sports program featuring the Indianapolis Colts
Classic Rock	"70's"	A classic rock format featuring music from the 1970's
Talk	Limbaugh	A call-in based talk show hosted by personality Rush Limbaugh

Table 9-3   Examples of the Use of the Program Type Name Feature

tailors the playlist and programs to best meet their listeners' needs. While a receiver may stop on several "Rock" stations, for instance, the listener can easily identify the station with a "70s" format when the PTYN feature is utilized.

Rules for Use of the PTYN Feature

To best implement the PTYN feature, the following rules should be followed:

1. The PTYN information is carried in group type 10A. This group type must be inserted into the group structure when transmitting PTYN information.
2. Up to eight alphanumeric characters may be used to describe the program content.
3. The PTYN is used to describe the program content in a general nature when the defined PTY is not adequate. The PTYN should not be used to transmit information of a detailed nature, such as artist and song title.
4. Since the PTYN is a semidynamic feature, it should follow changes to the broadcast PTY code. When changing from a program with a defined PTYN to one with no PTYN, then group type 10A should be removed from the group structure.
5. Never broadcast a PTYN containing all blanks.
6. All text segments of the PTYN should be transmitted. Four type 10A groups are required to transmit all eight characters.
7. The text A/B flag shall be toggled whenever the PTYN information is changed. This allows receivers to clear the PTYN text buffer and prevent a mixture of the old and new PTYN in the receiver display, especially under adverse reception conditions.

8. When changing PTYN information, it may be helpful to increase group type 10A repetition to quickly update receiver displays.

How the PTYN Feature Is Utilized in Receivers

As the PTYN feature is designed to augment the PTY feature, consumer receivers will replace the PTY information in the display with the PTYN information, if received. For example, a listener tunes to a station broadcasting the "Classic Rock" PTY code, and a PTYN of "70s." The receiver will display "Classic Rock" as soon as the station is tuned. After the PTYN information is adequately received, "70s" will be displayed. Most automotive receivers will replace the PTY portion of the display with the received PTYN. Since PTY information is transmitted in every RDS data group, the PTY code will be quickly received. A minimum of four type 10A groups must be received before the PTYN will be displayed. This will always result in a delay of display of the PTYN information in comparison to the PTY. The actual length of the delay will be dependent on the group structure and reception conditions, as well as the exact software implementation in the receiver.

An example of what a listener may see when performing a PTY seek is shown in Table 9.4. Note that the stations with no RDS data are never tuned to by the receiver, even though the desired program is being broadcast. Also

Frequency	Actual program content	Broadcast PTY	Broadcast PTYN	Receiver Display
89.7	Alternative Rock cut	Rock	Altrntve	Altrntve
92.5	Heavy Metal rock cut	Non RDS station	Non RDS station	Station is not tuned
94.7	A commercial	Rock	Not used	Rock
100.3	Heavy Metal rock cut	Rock	H. Metal	H. Metal
101.9	A currently popular rock cut	Non RDS station	Non RDS station	Station is not tuned
102.5	An acid rock cut	Rock	Acid	Acid
104.7	The local news	Rock	Not used	Rock

Table 9–4 Example of a Program Type Scan for "Rock"

of significance is that in the case of two stations that are not currently broadcasting music, but are airing a commercial and the news, the listener has still learned that the station offers his selected program type. Without RDS, the listener would most likely tune to another station, since the music program content is unknown. By using RDS, there is at least a chance that the listener will remain tuned throughout the commercial. The chances of this can be greatly augmented by utilizing the radiotext feature and transmitting "next up" information during the commercial.

Definition of the Terms Used to Denote North American Program Types

1	**News**	News reports, either local or network in origin.
2	**Information**	Programming that is intended to impart advice.
3	**Sports**	Sports reporting, commentary, and/or live event coverage, either local or network in origin.
4	**Talk**	Call-in and/or interview talk shows, either local or national in origin.
5	**Rock**	Album cuts.
6	**Classic Rock**	Rock-oriented oldies, often mixed with hit oldies, from a decade or more ago.
7	**Adult Hits**	An up-tempo contemporary hits format with no hard rock and no rap.
8	**Soft Rock**	Album cuts with a generally soft tempo.
9	**Top 40**	Current hits, often encompassing a variety of rock styles.
10	**Country**	Country music, including contemporary and traditional styles.
11	**Oldies**	Popular music, usually rock, with 80% or more being noncurrent music.
12	**Soft**	A cross between adult hits and classical, primarily noncurrent soft-rock originals.
13	**Nostalgia**	Big-band music.
14	**Jazz**	Mostly instrumental, includes both traditional jazz and more modern "smooth jazz."
15	**Classical**	Mostly instrumental, and usually orchestrations of adult hits.
16	**Rhythm and Blues**	A wide range of musical styles, often called "urban contemporary."
17	**Soft Rhythm and Blues**	Rhythm and blues with a generally soft tempo.
18	**Language**	Any programming format in a language other than English.
19	**Religious Music**	Music programming with religious lyrics.

20	**Religious Talk**	Call-in shows, interview programs, etc. with a religious theme.
21	**Personality**	A radio show where the on-air personality is the main attraction.
22	**Public**	Programming that is supported by listeners and/or corporate sponsors instead of advertising.
23	**College**	Programming produced by a college or university radio station.
24-28	**Unassigned**	
29	**Weather**	Weather forecasts or bulletins that are non-emergency in nature.
30	**Emergency Test**	Broadcast when testing emergency broadcast equipment or receivers. Not intended for searching or dynamic switching for consumer receivers.
31	**Emergency**	Emergency announcement made under exceptional circumstances to give warning of events causing danger of a general nature. Not to be used for searching; only used in a receiver for dynamic switching.

Definition of the Terms Used to Denote European Program Types

1	**News**	Short accounts of facts, events, and publicly expressed views, reportage, and actuality.
2	**Current Affairs**	Topical program expanding or enlarging upon the news, generally in different presentation style or concept, including debate or analysis.
3	**Information**	Program the purpose of which is to impart advice in the widest sense.
4	**Sport**	Program concerned with any aspect of sport.
5	**Education**	Program intended primarily to educate, of which the formal element is fundamental.
6	**Drama**	All radio plays and serials.
7	**Culture**	Programs concerned with any aspect of national or regional culture, including language, theater, etc.
8	**Science**	Programs about the natural sciences and technology.
9	**Varied**	Used for mainly speech-based programs, usually of a light-entertainment nature, not covered by other categories. Examples include: quizzes, panel games, personality interviews.

10	**Pop**	Commercial music, which would generally be considered to be of current popular appeal, often featuring in current or recent record sales charts.
11	**Rock**	Contemporary modern music, usually written and performed by young musicians.
12	**M.O.R.**	Middle of the Road Music. Common term to describe music considered to be "easy-listening," as opposed to Pop, Rock, or Classical. Music in this category is often, but not always, vocal, and usually of short duration.
13	**Light Classics**	Classical Musical for general, rather than specialist, appreciation. Examples of music in this category are instrumental music and vocal or choral works.
14	**Serious Classics**	Performances of major orchestral works, symphonies, chamber music, etc., including Grand Opera.
15	**Other Music**	Musical styles not fitting into any of the other categories. Particularly used for specialist music of which Rhythm & Blues and Reggae are examples.
16	**Weather**	Weather reports, forecasts, and Meteorological information.
17	**Finance**	Stock Market reports, commerce, trading, etc.
18	**Children's Programs**	For programs targeted at a young audience, primarily for entertainment and interest, rather than where the objective is to educate.
19	**Social Affairs**	Programs about people and things that influence them individually or in groups. Includes: sociology, history, geography, psychology, and society.
20	**Religion**	Any aspect of beliefs and faiths, involving a God or Gods, the nature of existence and ethics.
21	**Phone-In**	Involving members of the public expressing their views either by phone or at a public forum.
22	**Travel**	Features and programs concerned with travel to near and far destinations, package tours, and travel ideas and opportunities. **Not for use for *Announcements* about problems, delays, or roadworks affecting immediate travel where TP/TA should be used.**

23 **Leisure** Programs concerned with recreational activities in which the listener might participate. Examples include, Gardening, Fishing, Antique Collecting, Cooking, Food and Wine, etc.

24 **Jazz Music** Polyphonic, syncopated music characterized by improvisation.

25 **Country Music** Songs that originate from, or continue the musical tradition of, the American Southern States. Characterized by a straightforward melody and narrative story line.

26 **National Music** Current Popular Music of the Nation or Region in that country's language, as opposed to International "Pop," which is usually U.S. or U.K. inspired and in English.

27 **Oldies Music** Music from the so-called "golden age" of popular music.

28 **Folk Music** Music that has its roots in the musical culture of a particular nation, usually played on acoustic instruments. The narrative or story may be based on historical events or people.

29 **Documentary** Program concerned with factual matters, presented in an investigative style.

30 **Alarm Test** Not to be used for searching; only used in a receiver for dynamic switching.

31 **Alarm** Emergency announcement made under exceptional circumstances to give warning of events causing danger of a general nature. Not to be used for searching; only used in a receiver for dynamic switching.

10

Radiotext Transmission

One of the most powerful features of RDS is the ability to transmit text information to your listener. While the amount of text you may send is limited, it is the type of information you send that is much more important. As with any other RDS feature, it is important to orient the information to the current audio program. During music play, radiotext could be used to supply information such as:

1. Current artist and song title, and possibly release title:
   - "The Spirit of the Radio" by Rush
   - "And You and I" by Yes from *Closer To the Edge*
2. Next-up information to keep your listeners tuned in. This could be the next song or the next program, such as news or sports.
3. Between-tracks information, such as the current forecast, temperature, or station promotion could be given.

Radiotext allows you to send information that the listener may want without taking up valuable air time in the process. Some applications could include:

1. Next program, including on-air personality, news, sports, or special programs.
2. During commercials you may provide a variety of information about your advertiser. This should include information not normally carried in the audio portion of the commercial, such as store locations, hours, and telephone number.

3. During the news or sports, you could advertise special community events.
4. Under special circumstances, such as severe weather or traffic conditions, radiotext would allow you to give constant updates to those just tuning in.

How Radiotext Is Transmitted

Radiotext information is carried in either group type 2A or 2B, as shown in Figures 10.1 and 10.2. In a 2A group, up to sixty-four characters may be transmitted. A 2B group can carry up to thirty-two characters. As will be covered in Chapter 17, a "B" type group will transmit the PI code two times per group transmission, while a "A" type group transmits the PI code one time per transmission.

Due to the differing structure of the 2A/2B groups, four characters per group will be transmitted in a 2A group, while two characters per group will be transmitted in a 2B group. The character locations are denoted by the value of the text-segment address code. This text-segment address code can range from a value of 0 to 15 decimal. In this manner, the exact character location can be determined for each group transmission. Figure 10.3 shows a comparison of the two group types transmitting the text message "Welcome to WKAR Flint, Michigan." By knowing the character locations, receivers can provide for the error-correction of missed characters when the text message is repeated.

In this example, all text-segment addresses were required in the 2B group to complete the message, while only addresses 0-8 were required in the 2A group. It is not necessary to transmit addresses 9-15 in this example, since no information is carried. The message can be repeated beginning at text segment address 0, allowing best use of the data capacity. This also ensures that, over a given period of time, the text message will be repeated as many times as possible. The actual length of time required to complete a given text message is dependent upon the type of group used (2A or 2B), the number of text-segment addresses required, and the group structure being transmitted. The following example calculates the time required to transmit a sixty-four-character message given the following:

1. A group structure of 0B, 2A, 10A, 15B.
2. The length of time to transmit one group = 87.6ms.
3. The length of time required to transmit the given group structure = 4 × 87.6ms = 350.4ms.
4. The number of 2A group transmissions required for a 64-character message = 64/4 = 16.

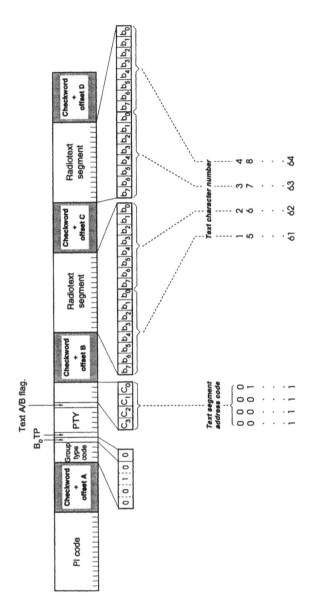

Figure 10–1 2A Group: Sixty-four-character Radiotext.

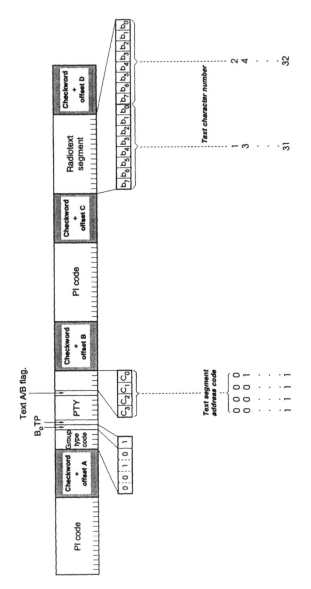

Figure 10–2 Group 2B: Thirty-two-character Radiotext.

Text Segment Address	2A Group	2B Group
0	Welc	We
1	ome_	lc
2	To_W	om
3	KAR_	e_
4	Flin	To
5	t,_M	_W
6	ichi	KA
7	gan.	R_
8		Fl
9		in
10		t,
11		Mi
12		ch
13		ig
14		an
15		._

Figure 10–3 Comparison of radiotext transmission in 2A versus 2B groups.

5. The total length of time to complete the text message = 16 × 350.4ms = 5.6064s.

The exact same time will be required to transmit a thirty-two-character message using the 2B group. The length of time required to send a text message can be given by the generic formula:

$$\text{RT time} = \frac{\text{(Number of groups in group structure) (TSA) (87.6ms)}}{\text{Number of type 2 groups in group structure}}$$

Where: TSA = $\dfrac{\text{Number of characters in message}}{\text{4 for 2A groups or 2 for 2B groups}}$

This assumes that no unused text-segment addresses are transmitted. If the encoder in use transmits all text-segment addresses, regardless of the number of characters in the message, then the TSA will always be equal to 16. It can be seen that increasing the repetition of the radiotext groups will greatly reduce the amount of time to complete the message.

The Text A/B Flag

The text A/B flag is an important part of proper radiotext transmission. This flag is used to signal the receiver when a new text message is transmitted. The bit should toggle whenever the text information is changed. Thus, if the A/B flag was a "0" in the prior message then it should change to a "1" for the next message and vice versa, as shown in Figure 10.4.

When the receiver detects a change in the A/B flag state, the radiotext receiver buffer will be cleared, preventing the possibility of a mixture of old and new text messages being displayed on the receiver. Depending upon the exact implementation of the radiotext feature on the receiver, it may still be possible that an incomplete message is displayed, but by properly using the A/B flag, mixed text messages can be prevented.

How Fast Can Text Messages Be Changed?

While a complete text message can be transmitted rather rapidly, the main reason in doing so is to insure proper reception of the data, rather than to

A/B Flag	Radiotext message
0	Now playing on WKRQ - Genesis And the lamb lies down on broadway
1	WKRQ plays the best from the 60's 70's and 80's
0	Next up - Van Halen Jamies Crying on WKRQ
1	The news is brought to you by Engles Jewelers. Phone 356-1100!

Figure 10–4   Example of text A/B Flag Operation.

simply change the text message again. The ability for a receiver to properly receive and display a message is dependent upon the following:

1. Number of characters in the message.
2. Repetition rate of the 2A or 2B data group.
3. Reception conditions at the receiver.
4. Software implementation of the radiotext feature.
5. The user accessing the feature on the receiver.

In the above list, the most significant factor is the listener accessing the feature. When you account for the fact that a large number of users will be in a car, then there is simply no way that constantly changing text would be of any advantage. The text data should also be analogous to the current program material as well, making the necessity of constant updates unnecessary. Some general guidelines for radiotext transmission are:

- Keep the data in line with the audio program content; this provides the maximum usefulness to the listener.
- Remember that the user must access the feature; if it is changed too rapidly, then many messages will be overlooked.
- Reception conditions vary widely; repeat the text data for no less than 30 seconds to ensure proper reception.
- Ensure that unused text-segment addresses are not transmitted in order to improve data efficiency.
- If multiple type 2 groups are specified in the group structure, spread them apart from one another to improve time diversity (i.e., use 0A, 2A, 15A, 2A, 10A instead of 0A, 2A, 2A, 15A, 10A).
- Do not mix the transmission of type 2A and 2B groups; this can cause the receiver to decode no text data.
- The text A/B flag should be toggled whenever a new text message is transmitted.

By keeping these rules in mind, you will afford the listener the greatest usefulness and versatility of this powerful feature.

11

Considerations When Transmitting Text

The ability to transmit textual information to a consumer receiver is very powerful. Using the data transmission methods available you can also transmit free-form text to specialized data receivers. The predefined consumer-oriented RDS features that allow text transmission are:

- **Program Service (PS) Name.** An eight-character name for your station. This information is static.
- **Program Type Name (PTYN).** An eight-character description that provides additional detail to the defined Program Type (PTY) code. This information is semi-static, as it is changed only with changes to the program content.
- **Radiotext (RT).** Thirty-two- or sixty-four-character messages containing any type of information you would like to send to your listener, such as current artist and song title, next-up, or station promotional messages. This information may be static, semi-static, or dynamic in nature.

Before you begin transmitting text, you will need to consider your audience. Listeners will be using a variety of receiving equipment with differing display sizes and capabilities; they will also be performing a variety of tasks, such as walking, driving, or listening at home. The listener's ability to make use of text information is a combination of these two factors. While it may seem that you cannot account for all the people all of the time, your task can be made much simpler by remembering the underlying purpose of text data, which is to impart useful information about the current program. "Program" refers to the current audio broadcast, whatever it may be. By thinking of the text features as extensions of the audio broadcast, rather than non-associated data, you will find that the content, repetition rate, and dynamics of each feature will flow in the same way that your audio program flows. Rather than approaching these new features as "techno-gadgets," approach them for their usefulness in imparting vital program information quickly and inaudibly to the listener.

On a practical level, we may now approach the listener more closely by looking at the various types of displays that may be available. Displays may be characterized by two main criteria:

1. *Display size.* The total number of characters that may be presented in one display. The minimum display size for an RDS receiver is one row of eight characters.
2. *Character generation.* A combination between the character style and the character ROM. The two most common types are British flag and dot matrix.

Display Size

The minimum RDS display size is eight characters. This is due to the fact that the PS and PTYN features are defined as eight characters in length. Since the PS will be displayed for the majority of the time in place of the frequency, then all RDS receivers will have at least eight characters. After meeting this minimum criteria, then considerations of cost and physical size become important. The larger the display, the more expensive it will be. Since only one RDS feature, radiotext (RT), has more than eight characters, increasing the display size over eight characters has limited benefit. If the decision is made to have a display longer than eight characters, often additional RDS and non-RDS information can be displayed simultaneously when the RT is not currently being viewed. The obvious issue is how to display messages longer than the display size in a fashion that is easy to read, given the application. Here, two basic methods may be employed:

1. *Text scrolling.* Text is moved right to left across the display one character at a time at a rate that allows the user to view the entire message.

Advantages	**Disadvantages**
Least costly, since only eight characters are required.	Harder to read and interpret.
Requires minimal software to implement.	Requires the viewer's constant attention as the text "crawls" across the display.

Requires less physical space, which is at a premium on portable or mobile applications.

Could be considered more crude an implementation.

2. *Text formatting.* Text is presented one full display at a time. Words that will be broken are wrapped into the next display. This limits broken words to only those that are longer than the total display length.

Advantages

Easier to read and interpret.

Does not require the viewer's constant attention. Text rate can be controlled by the viewer.

More sophisticated presentation of text.

Disadvantages

More costly, since larger display sizes are required.

Requires smarter software to implement.

Requires more physical space on the receiver.

Each method has its advantages and disadvantages and should be carefully weighed against the intended application.

Character Generation

The ability to represent characters on a receiver display is dependent upon two basic criteria:

1. The type of display being either a fourteen- or sixteen-segment British flag, or a 5×7 dot matrix. The dot matrix display has the ability to represent many more characters than the British flag but is more expensive.
2. The amount of available character ROM. The dot matrix display may have the ability to form many characters, but the display driver IC must have enough memory to store all the representations in ROM. The amount of character ROM available is limited due to cost restraints.

Display Types

The next consideration should be the selection of characters that you use. Table E.1 of Annex E of the *Standard* is a table displaying all the possible characters that can be transmitted. While any one of these characters may be transmitted, it must be displayed on the receiver to be of use. As mentioned earlier, most consumer receivers will use either a fourteen- or sixteen-segment British flag or a 5 × 7 dot matrix display. The lower-cost British flag display can only display a portion of this table. Receivers with this type of display will often remap lower-case alphabetical characters that cannot be displayed as upper-case characters. If a received character is not displayable by the receiver, then it will be represented by a space in that character position. The following characters may be considered to be displayable on all receivers:

1. All upper- and lower-case alphabetical characters.
2. Numbers.
3. Normal punctuation, such as periods, hyphens, commas, and apostrophes.

Whatever display type is used, it must be at least eight characters in length. Typical display types are shown in Figures 11.1 and 11.2.

Figure 11–1 A sixteen-segment British flag type display.

Figure 11–2 A 5 × 7 dot matrix display.

12

Implementing Traffic Services through RDS

Overview

A popular feature of RDS is the ability to supply traffic information to your listeners. You probably already do this today in some form or another. You may even pay a traffic information service provider as well as obtain traffic information sponsors with some of your advertising clients. Listeners tune to your station for the latest in traffic information during the drive to and from work. There may be some listeners who tune to your station simply because you offer the best traffic services in town. You may even gain listeners because they have tuned to your station specifically for traffic information.

With the RDS traffic program feature, you can attract new listeners whose radios automatically tune to your station when they activate the traffic program feature. These listeners will always hear the sponsored advertisement that precedes the latest traffic bulletin, even if they are listening to their CD or have turned the volume all the way down. If this sounds intriguing, read on to see just how easy it is to do this.

One Bit, Two Bits, That's All It Takes

The RDS traffic feature can be implemented in two simple bits of the RDS data stream, as shown in Figure 12.1. Have you ever wished that you could have a remote controlled switch on your listener's receiver so that he would listen to what you have to say? The traffic feature does just that. The two bits related to the traffic feature are:

Traffic Program code (TP)	Traffic Announcement code (TA)	Applications
0	0	This program does not carry traffic announcements nor does it refer, via EON, to a program that does.
0	1	This program carries EON information about another program which gives traffic information.
1	0	This program carries traffic announcements but none are being broadcast at present and may also carry EON information about other traffic announcements.
1	1	A traffic announcement is being broadcast on this program at present.

Figure 12-1   Traffic Program and Announcement Codes.

- *Traffic Program (TP)*. When this bit is set (Logical 1), it indicates that your station gives traffic bulletins. RDS receivers with the traffic feature will search for this bit as it searches for traffic programs when the user activates the feature.
- *Traffic Announcement (TA)*. This bit should be set high (Logical 1) at the beginning of a traffic bulletin. When an RDS receiver detects this bit set, it will switch over to your broadcast. At the end of the traffic announcement, this bit should be set low (Logical 0). The receiver will return to its prior mode at that point.

As you can see, the traffic feature can be implemented with a simple single-pole single-throw switch if you desire. This "switch" changes the mode of all the RDS receivers tuned to your station that have the traffic feature turned on.

The Traffic Program (TP) Bit

The first rule is simple: do not use this feature if you do not provide the service. Even though it may be tempting to gain listeners by "saying" you offer traffic programs, you will quickly alienate listeners when they figure out that you do not. The goal of offering this or any RDS service is to attract listeners who have indicated by the feature settings on their receiver that they want this information. Listeners will turn against unfair advertising of RDS features, so resist the temptation. If you do not offer traffic programming as described below, then the TP and TA bits should be set low (logical 0).

We have now determined that you want to offer traffic programming through RDS. The TP bit should be set high. Most often you will give traffic bulletins during the morning and afternoon rush periods. You may provide traffic bulletins during the rush period on regular intervals—every fifteen minutes, for instance—to insure that everyone has the latest information as they begin their commute. Traffic bulletins at other times will probably only be given under severe circumstances. It is acceptable to leave the TP bit set high twenty-four hours a day under these conditions.

If your station provides traffic bulletins rarely, or only in extreme circumstances such as a major shutdown or blockage, then the TP bit should normally remain set low. Prior to issuing a traffic bulletin, the TP bit should be set high. After the bulletin is given, the TP bit should be set low again if you do not expect to give another update within a reasonable amount of time.

The Traffic Announcement (TA) Bit

The use of the TA bit is quite simple. This bit should be set high at the beginning of a traffic announcement, then set low again at the end of the announcement. This signals the receiver to switch to the FM tuner mode. Many RDS receivers allow the listener to set a default volume level for the traffic announcement so that the volume will be set to a consistent level during announcements. Figure 12.2 represents the TP and TA code usage during an announcement.

Sponsored Traffic

If your traffic announcements are sponsored, it is acceptable to begin or end each traffic announcement with the sponsor advertisement. This sponsorship should be kept short in relationship to the overall announcement. Under no circumstances is it acceptable to air a full-length ad while the TA bit is set high. Some acceptable messages are:

- WZPL traffic, brought to you by Eric's Chevrolet. Come by and see the end-of-year close-out bargains at Eric's.
- This traffic update was sponsored by the good folks at Karma Records. This week's special is the new release by Rush, "Test for Echo," now available for only $12.99.

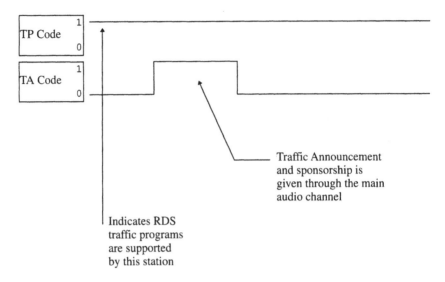

Figure 12-2 A Typical non-EON Traffic Announcement.

As you may imagine, there is great power in supplying advertising to listeners who may have been playing a CD. Take advantage of this with your sponsors, but do not abuse it. The key to properly using the RDS features is to meld them in with your audio programming. The RDS data becomes an extension of your audio program. Let the audio drive the data rather than letting the data drive the audio.

Setting the TP and TA Bits on the RDS Encoder

Most encoders include a software interface program that allows you to set default conditions, such as the Program Service (PS) name, Program Identification (PI) code, and so on. This is where the TP and TA bits may also be set. In most cases, the TP code will be set low or high and will remain in that state. The TA code, however, must be switched dynamically by some means. Some possible methods of setting the TA code are:

- *With a hardware "TA" switch.* Some encoders have an interface that allows a single switch closure to set the TA bit high or low. With this configuration, someone must close the switch at the beginning of the traffic announcement, then open the switch at the end of the announcement.
- *With a remote "Data Record" selector.* Many encoders allow the storage and retrieval of multiple data records. A data record contains all the essential RDS programming. If you define two records that are

identical, with the exception of the TA code being low in one and high in the other, then you may select the appropriate record to change the TA status. Data records may be selectable through either software control or even a hardware switch. You can use additional RDS features, such as radiotext, the music/speech flag, and decoder information to augment the announcement as well.

- *Through subaudible/audible tones.* Many commercial devices are available for the encoding and decoding of subaudible and audible tones. It is possible to encode these tones with the traffic sponsor spot, or even with a jingle played at the beginning and ending of the traffic message. A tone decoder with a relay output supplied with audio from a conventional receiver can then be used at the encoder location to switch the TA code or data record.
- *Full automation systems.* Planned traffic announcements can be programmed into some music or playlist automation systems. The automation system may then drive the RDS encoder directly through a direct or remote RS-232 interface. Speak directly with your automation system provider if you are unsure of its RDS interface capabilities. Remember that your system provider will respond to market demands; it is, however, up to you to let your requirements be known.

In all instances, it is important that the TA code be set low as soon as the traffic announcement is given. If a relay system is designed, it should contain a timer interface that will automatically set the TA code low in the event that the automation system fails to open the relay.

Remote Traffic Feeds

If your traffic information is supplied through a commercial service, such as Shadow or Metro traffic, you may consider negotiating with your service provider to provide the RDS control links necessary for setting the TA code. Many of the means described earlier for control of the TA code can be utilized in remote feeds. Dial-up modems or audible or subaudible tones are good ways to give limited access to the service provider; they also leave you with one less worry.

Sharing Traffic Information through the Enhanced Other Network Feature

The Enhanced Other Network or EON feature allows you to provide traffic information to your listeners without actually carrying the traffic announcement audio on your station. This may sound a bit bizarre, but this feature is commonplace throughout Europe. The EON feature supplies information through the RDS datastream about other affiliated stations. Don't worry about receivers being switched to your competitors station; the only way the receiver will learn anything or switch to another station is if you tell it to do so through the RDS EON data that you transmit. So relax, and lets find out how to save time and money with this useful feature.

EON Traffic: A Case Scenario

In this example we will assume that we are station KCBS. We are affiliated through ownership with KKID. Both stations are in the same market and have a large portion of overlapping coverage. The two stations have different formats and are therefore not considered competitive. In fact, KCBS is a classical format, while KKID is a rock format. KKID buys traffic information through a commercial service and supports the RDS traffic feature.

What you would like to do is provide your listeners with the latest traffic information, but you do not have the resources to obtain the commercial service. Furthermore, you really don't want to interrupt your classical format with a bunch of traffic announcements. In fact, you are worried that you may alienate some listeners if you carry traffic. These listeners may tune to your competitors' classic station KABC if you deluge them with a bunch of traffic bulletins. What do you do?

To solve this problem, you must give your listeners who want traffic bulletins the information they desire, without interrupting your audio program for the listeners who don't. You can do this by incorporating the RDS EON feature. It will require some coordination between you and your affiliate, but remember, you don't want your listeners tuning away to either obtain or avoid the traffic information. RDS EON traffic will solve this dilemma by tuning your listeners from your station KCBS to your affiliate station KKID at the beginning of the traffic bulletin. At the end of the bulletin, the receiver will return to your station. Only those listeners who have turned on the traffic announcement feature on their radios will hear the bulletin. Those without RDS radios, or who have the traffic feature disabled on their radios, will stay tuned to your station without interruption of the program. Problem solved!

Setting Up the RDS EON Feature

As shown in Figure 12.1, setting TP low and TA high on your station identifies that you provide traffic information through the EON feature. You must enter these settings into your encoder. Additional data required for this feature is carried through data groups 14A and 14B. Group 14A is used to supply information about your affiliate's station, such as the program identification (PI) code, program service (PS) name, and the frequency (or frequencies) that the station is available on. This information is stored in the receiver to allow instant tuning and recognition of your affiliate's station. This data must be entered into your RDS encoder's EON information. You must include group type 14A in your group structure if you are supplying EON information. It is not necessary to repeat this data as frequently as other data, but all EON information must be repeated at least every two minutes. When your affiliate is ready to begin a traffic announcement, your encoder must transmit a burst of group type 14B. This group tells the receivers to tune to your affiliate's station for the traffic announcement. You must ensure that your affiliate has set the TA code high on his station within two seconds of your transmitting the 14B group. Upon receipt of the 14B group the RDS receiver will:

1. Tune to each of the frequencies listed for the affiliated station and check that the signal quality is acceptable. If the station is not strong enough, the receiver will retune to your station.
2. Verify that it is indeed tuned to the correct station by verifying the received PI code against the one stored in memory. If it is the wrong station, the receiver will retune to your station.
3. Check the status of the TP and TA codes. If the TP and TA codes are not set high within two seconds, the receiver will retune to your station.
4. Unmute the audio and set it to a user preset level. The listener now hears the traffic announcement.
5. When the receiver detects that the announcement is over by a change of the TA code back to low, the receiver will retune to your station.

The EON traffic message will be ignored if the listener has the traffic feature off or the receiver does not have the EON feature. Receivers capable of this feature will have a special RDS EON logo on the front of the receiver. Just because a receiver has the traffic feature does not mean that it is EON capable; it must also carry the EON logo. Figure 12.3 depicts the EON traffic feature operation.

Automation Requirements for EON Traffic

To supply EON traffic information successfully some means of coordinating the RDS data of both stations to the audible announcement must take place. Let's review the data requirements at the two stations, as shown in Figure 12.3.

The key is to coordinate the switch from station 1 to station 2 at the beginning of the traffic announcement. The only dynamic data that station 1 must supply is the burst of the 14B data group just as the TA code of station 2 is set high. Some possible means of doing this are:

RDS Data	Station 1 Supplies EON Traffic	Station 2 Supplies non-EON Traffic	Purpose
TP, TA	TP=0, TA=1 Static data	TP=1 TA=0 except during traffic announcement	Denotes EON traffic on Station 1, denotes non-EON traffic on Station 2. Returns the receiver to Station 1 at the end of the announcement
Group 14A	14A Supplies EON data for Station 2 Static data	N/A	Gives the receiver the PI,PS,AF, and traffic settings for Station 2
Group 14B	Burst of 14B's when TA of Station 2 is first set to 1	N/A	Tells the receiver to switch now to Station 2 for traffic information

Figure 12-3   Data requirements for EON traffic.

- *An automated data link from station 2 to station 1.* This would possibly be through a computer type interface that sends the proper RS-232 control commands to the encoder of station 1 as the TA code of station 2 is set high.
- *Subaudible or audible tone decoder.* A receiver, tuned to station 2 and placed at station 1, that would decode a tone that is transmitted on station 2's audio when the TA bit is set high and execute a relay closure or command sequence to the encoder at station 1.
- *RDS data receiver that will monitor the status of the TA code.* An RDS data receiver tuned to station 2 and placed at station 1 that will monitor the status of the TA code. When the TA code is changed from low to high, either a relay closure or command sequence is issued to the encoder of station 1.

Providing Localized Traffic Announcements

One benefit of providing EON traffic is the ability to provide localized traffic information. We will again explore this possibility through an example. In this example we have a large metropolitan area known as Megalopolis. Megalopolis radio station WMGP provides coverage throughout the city but does not extend into the suburbs of Northtown, Southtown, and Westtown. Each of the suburbs has a small local station with coverage extending partly into Megalopolis. Many commuters travel to and from the city to the suburbs as part of their daily commute.

WMGP would like to provide traffic information to these commuters. The manager at WMGP tried for some time to carry non-EON traffic announcements, but found that too much air time was being consumed, because they had to provide traffic information for commuters from Northtown, Southtown, and Westtown. Additionally, many listeners were annoyed at having to listen to all those traffic announcements for areas they had no interest in.

The engineer at WMGP devised a brilliant plan to employ the use of EON traffic announcements. Through EON, WMGP had only to supply the necessary RDS data silently to their listeners. Specifically, data group 14A was set to provide the required tuning information for WNST, WSST, and WWST. Then, when any of these stations began a non-EON traffic announcement, WMGP simply had to transmit a 14B data group with the proper Program Identification (PI) code for the station(s) beginning an announcement. The listeners tuned to WMGP will only hear a traffic announcement when

1. their RDS EON-equipped receivers have the traffic feature on, and
2. they are within the coverage area of the station carrying the announcement.

In this case, driver A will hear only announcements for WNST. Driver B will receive no traffic announcements. Driver C will hear only announcements for station WSST. Our problem scenario has been solved through the use of EON. While the solution does involve cooperation between all four stations, since the RDS data for initiating the traffic announcements must be coordinated, it does illustrate that listeners can never be tuned to a different station without your direct involvement. The basic data requirements are summarized in Figure 12.5.

Providing Local and Localized Traffic Information via EON

In our example, station WMGP desired to eliminate all traffic announcements from its audio program. Driver B received no traffic information. In

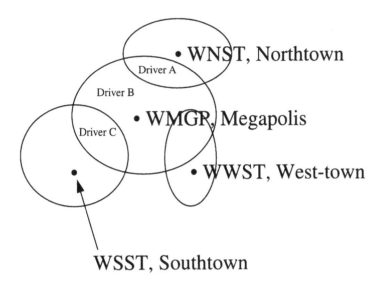

WSST, Southtown

Figure 12-4 Localized traffic through Enhanced Other Network.

Station	TP	TA	EON data via group 14A	EON data via group 14B
WMGP	0	1	Provides PI,PS, TP, TA, PTY, and PIN data for WSST, WWST, and WNST	Transmits a burst of 14B groups for the proper station at the beginning of a traffic announcement
WSST	1	0 except during a traffic announcement	N/A	N/A
WWST	1	0 except during a traffic announcement	N/A	N/A
WNST	1	0 except during a traffic announcement	N/A	N/A

Figure 12-5 RDS EON-related data for localized traffic.

the case of most big cities, traffic in the downtown area can often be a problem. Our case scenario can be modified just a bit to allow traffic announcements for the WMGP city proper listening area, and still provide localized traffic information in the suburban listening areas through stations WSST, WWST, and WNST. All that is required is that WMGP set its TP code high during the time that it carries a local traffic announcement through its audio. In this situation, all listeners within WMGP's listening area, including driver B, will receive the traffic announcement. The amount of time devoted to audible traffic announcements on WMGP is greatly minimized by combining the features of local and EON traffic announcements.

13

Alternate Frequency Feature

One of the most powerful features of RDS is the ability to automatically link the receiver to two or more transmitters carrying the same program material. The receiver will automatically switch from one transmitter to another when the signal quality becomes unacceptable. There is no risk of someone stealing listeners away, since the receiver requires the proper verification before switching to an alternate frequency. The Alternate Frequency feature or AF makes use of three other RDS features to make the operation seamless to the listener. The AF feature works as follows:

1. When tuned to a station, the receiver obtains AF information through group type 0A.
2. The received AF information is stored with the received Program Identification (PI) code in memory.
3. The received Program Service (PS) name is displayed on the receiver in place of the frequency. The PS is also stored in the receiver memory.
4. When the received station's signal quality becomes unacceptable, the receiver will quickly measure the signal quality of the alternate frequencies.
5. The receiver will tune to the strongest AF available. If no AFs are acceptable, the receiver will remain tuned to the current station.
6. Once tuned to an AF, the receiver will verify that the same network has been tuned by matching the received PI code with the expected PI code.
7. If the PI codes match, the receiver remains tuned to the AF. If the PI codes do not match, the receiver will retune to the last frequency.

While this description of the AF feature is somewhat simplified, it covers the basic operation. In this example, the listener doesn't care that the frequency has been changed since the displayed PS has remained constant. The listener has the perception that the station's coverage area is much larger than it actually is. With the RDS AF feature, FM stations can obtain equivalent or even greater coverage areas than are possible on AM. In fact, with enough stations, it is quite possible to provide nationwide service without requiring the listener to ever retune the receiver.

Designing an AF Transmitter Network

To make the RDS AF feature work, the following requirements must be met:

1. All transmitters must be equipped with RDS.
2. All transmitters must transmit the same PI code.
3. Each transmitter must transmit an AF list that includes all the frequencies in the network.
4. The transmitters must have overlapping coverage areas.
5. All transmitters should transmit the same PS name, although it is possible to transmit different PS names without interrupting receiver operation.

Coverage Overlap

When designing a transmitter network, it is important that the transmitters have sufficient coverage overlap. The overlap must be based in reality rather than simply calculated values. The best method of achieving the proper overlap is to design the system taking into account physical and natural interference mechanisms, such as terrain and buildings. Areas prone to multipath interference must be designed to allow at least one unobstructed transmitter along the receive path. A general rule of thumb is to provide an average of at least 30 dBuV of overlap. Figure 13.1 depicts a well-designed transmitter network. Higher values up to 36 dBuV can provide some advantage, but amounts over that are not necessary. The goal is to provide the receiver with sufficient signal strength to enable a graceful switchover from one transmitter to the next and allow the listener to achieve uninterrupted listening. It is not acceptable for the listener to wait for any amount of time before the next transmitter is receivable. It is likely that once it is no longer possible to receive the network, the listener will select another station. Note that all four stations broadcast the same PI code. While the PI code is calculated from each station's call sign, one of the station's PI code must be chosen for use by all the transmitters in the network. Of course, this is a con-

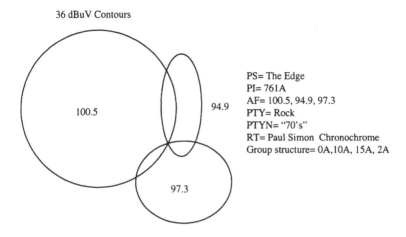

36 dBuV Contours

PS= The Edge
PI= 761A
AF= 100.5, 94.9, 97.3
PTY= Rock
PTYN= "70's"
RT= Paul Simon Chronochrome
Group structure= 0A,10A, 15A, 2A

Figure 13-1   A typical RDS Alternate Frequency network.

cern only where there is more than one full-power licensed transmitter in the network.

Transmitter Configuration

The number of RDS encoders required for an AF network will depend on the exact configuration and routing of audio and control signals. If analog Studio to Transmitter Links (STL) are employed throughout the network, then only one RDS encoder is required. Figure 13.2 depicts the network configuration for an analog STL system. With this configuration, the RDS data on all the transmitters will be identical. This should not be a problem, since all the transmitters carry the same audio. If digital STLs such as a T1 line are used, then an RDS encoder will be required at each transmitter site. Dynamic control of the encoders will have to be coordinated through some external means, such as an RS-232 interface. Depending on the RS-232 configuration, the encoders may be controlled simultaneously, if desired. Figure 13.3 depicts a digital STL network system.

Partial Simulcast Networks

Some large full-power networks share program audio, but differ during commercials. This is done in order to provide local traffic to each market while sharing the same audio program. This type of commercial application requires special attention for true compatibility with consumer receivers. During the period when advertising is aired, the audio content will differ. If

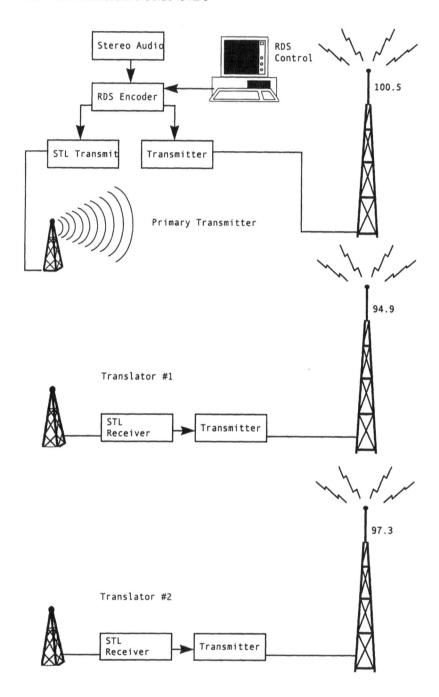

Figure 13–2 RDS Alternate Frequency Network utilizing analog studio to transmitter links.

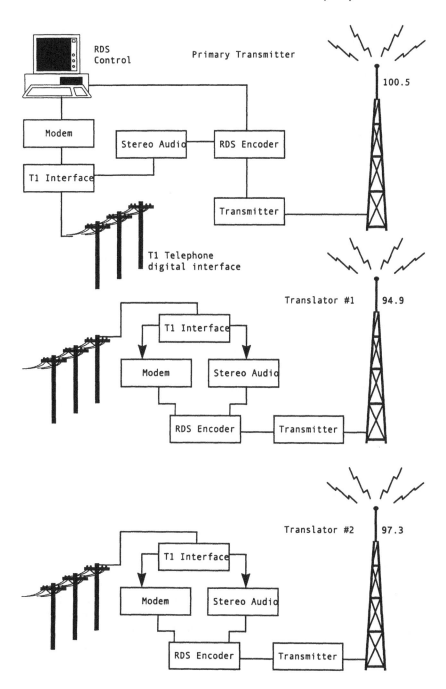

Figure 13–3 RDS Alternate Frequency Network utilizing digital studio to transmitter links.

a listener is driving at the switchover point between two transmitters, then it is possible that the radio will switch from one transmitter to the other. The listener perceives that something wrong has happened, since the audio program has suddenly changed. To examine this problem we will look at three different possible solutions.

Case 1: Static, Identical RDS Data

The simplest solution is to do nothing. In this scenario, all transmitters in the network are given as AFs and share the same PI code. If a listener is traveling in the transmitter transition area during a commercial, the receiver may switch from one transmitter to the next, causing a disruption in the audio continuity. This may be acceptable, since there will only be a limited number of listeners traveling through a transition period during a commercial period. In an environment with large flat areas, the receiver will likely switch only once from one transmitter to the next. If, however, the location is prone to mutipath or temporary blockages of RF, it is possible for the receiver to switch many times between two or more transmitters. In this case, the listener will probably switch to another station instead of being tuned back and forth between differing commercials. Since we cannot predict the response of the listeners to the receiver switching between nonsimulcast audio, we cannot judge their acceptance. Remember, the listener thinks he is tuned to the same station based on the received PS name that is displayed on the receiver, but the audio program has changed. The advantage of this setup is that a single RDS encoder may be used if analog STLs are employed. A slight variance that can be introduced will be examined next.

Case 2: Static RDS Data with Differing Program Service (PS) Names

Operation in Case 2 is identical to Case 1 with one exception. In Case 2, each transmitter is assigned a different PS name. This requires a separate RDS encoder at each transmitter site. The receiver can still switch to nonsimulcast audio during a commercial period, but the receiver will indicate a change by displaying the new PS name on the receiver. This offers the listener an explanation as to why the audio content is different, and may increase acceptance. This configuration also gives each transmitter its own unique identity in each market. Since each station is retaining its identity through the use of local advertising, then this application may be a good compromise except in areas of high multipath. One switchover in the transmitter transition zone is acceptable, but multiple switching between transmitters during commercial periods will likely result in the listener tuning to another station.

Case 3: Regional Variant PI Codes

In this case, the regional variant feature of RDS is utilized to prevent consumer receivers from switching during commercials. To accomplish this, the following is required:

- A network PI code must be obtained. In North America, special PI codes are assigned for networked-based operation. These special PI codes are interpreted differently by consumer receivers when performing an AF switch. A network PI code may be obtained from the National Association of Broadcasters. This PI code will be used on all the transmitters in the network.
- The PS names may be the same or different.
- A separate RDS encoder must be placed at each transmitter. Each encoder must be dynamically controllable through some external means. This control link is required to change the PI code during the commercial period.

Utilizing the Regional Variant Feature

For our example, we will assume that a regional variant PI of Bx12 was assigned for our networks use. As was discussed in Chapter 7 regarding program identification (PI) codes, the second nibble of the PI code carries important information about the network programs. As shown in Figure 13.4, variances in the second nibble convey important information to consumer receivers. During the periods when all the transmitters carry identical programming, the second nibble of the PI code should be set to 3, yielding a PI code of B312. The regional variant set to 3 identifies the broadcast as a supraregional network. Since our broadcast is neither international or national in scope, this is the proper application. During periods where the audio differs among the transmitters, each transmitter's PI code must be changed to a different regional variant. Regional variants are 4-F. It does not matter which of these variants you use, as long as they are all different. This operation is detailed in Figure 13.5.

The regional variant feature is implemented in consumer receivers. Many receivers will allow the listener to select whether or not they will accept or reject variant AF switches. Consumer receivers should implement the regional variant feature as follows:

- A switch should be provided to allow the consumer to accept or reject variant AF switches. The feature should default to the reject variant setting.

Area coverage code	L	I	N	S	R1	R2	R3	R4	R5	R6	R7	R8	R9	R10	R11	R12
HEX	0	1	2	3	4	5	6	7	8	9	A	B	C	D	E	F

Bits b_{11} to b_8:Second nibble or digit of the PI code

I: (International)   The same program is also transmitted in other countries.

N: (National)   The same program is transmitted throughout the country.

S: (Supra-regional)   The same program is transmitted throughout a large part of the country.

R1 . . . R12: (Regional)The program is available only in one location or region over one or more frequencies, and there exists no definition of its frontiers.

L: (Local)   Local program transmitted via a single transmitter only during the whole transmitting time.

Figure 13–4 Coverage area codes for nationally allocated program identification (PI) codes.

Frequency	PS Name	PI Code	Time	Program Content
100.5	WLRT-FM	B312	12:10	Network programming
94.9	WSHU-FM	B312		
97.3	WSDR-FM	B312	12:20	Receivers will AF switch
100.5	WLRT-FM	B412	12:21	Local commercials and news
94.9	WSHU-FM	B512		
97.3	WSDR-FM	B612	12:30	Receivers will not AF switch
100.5	WLRT-FM	B312	12:31	Network programming
94.9	WSHU-FM	B312		
97.3	WSDR-FM	B312	12:45	Receivers will AF switch
100.5	WLRT-FM	B412	12:46	Local commercials and weather
94.9	WSHU-FM	B512		
97.3	WSDR-FM	B612	12:50	Receivers will not AF switch
100.5	WLRT-FM	B312	12:51	Network programming
94.9	WSHU-FM	B312		
97.3	WSDR-FM	B312	1:00	Receivers will AF switch

Figure 13-5 Networked-based programming based on regional PI codes.

115

- If no switch is provided, the receiver should always reject variant AFs, except during the following conditions:
  During a preset recall
  During a PI seek

Utilizing Regional Variants with Loosely Affiliated Networks

The regional variant feature can also be used to provide enhanced tuning among loosely affiliated networks, such as National Public Radio (NPR) or even syndicated programs. This feature is intended for NPR more than other applications, however, as we will explore.

National Public Radio

NPR consists of a nationwide network of stations with membership in the NPR organization. NPR provides program material to all stations over a number of satellite channels. Many NPR affiliates make use of FM translators or other full-power stations that can take advantage of the Alternate Frequency (AF) feature of RDS. Most listeners know that just about any place they travel they will be able to find an NPR member station. The problem is that member stations are often individually run, with program material often differing from one "network" to the next. Member stations can provide a variety of programming, such as:

- Direct-feed satellite programs. There are a number of programs to choose from at any given time.
- Tape-delayed NPR programs.
- Local programs.
- Local news and weather information.

Based on the multitude of program options that any member station has at any given time, it is undesirable to link all NPR stations together by utilizing the same PI code at all times. However, NPR may make use of a nationally assigned PI code of Bx01. With some advance planning and coordination between neighboring stations, NPR member stations can provide the following features for their listeners without having to dynamically alter transmitted RDS data:

- Receivers will automatically tune within transmitters shared by the same member station and its affiliates. These networks carry the same program material at all times.
- Receivers will not AF switch from one member network to another.
- When the listener presses the receiver preset recall, which has been set to any NPR affiliate station, the receiver will automatically tune to

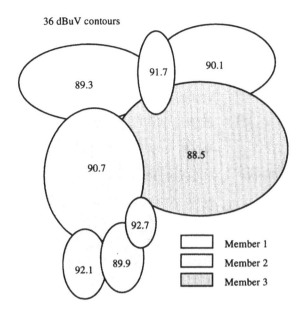

36 dBuV contours

Figure 13–6
National Public Radio Network.

the strongest available NPR station, regardless of which member network was last tuned.

- If no NPR affiliated stations are detected during the preset recall, the receiver will automatically tune across the band, attempting to tune to any NPR affiliate station.

To accomplish all the above features, each NPR station is required to utilize the nationally allocated PI code of Bx01. Additionally, each station should transmit information about the adjoining member stations through the Enhanced Other Network (EON) feature. For this example, we will use the NPR network depicted in Figure 13.6. This consists of three different NPR member stations and their affiliates. These stations are described as follows:

- Member 1 consists of two full-power stations and one translator. The audio content on all three transmitters is always the same.
- Member 2 consists of one full-power station and three translators. The audio content on all four transmitters is always the same.
- Member 3 consists of a single transmitter.

To properly make use of RDS, the stations must transmit the RDS data shown in Figure 13.7. In reality, NPR is much larger than this example. By simply adding more EON records for nearby NPR member networks, the entire country can be covered. The only rule is that the regional PI codes

Member	PS	PI	AF Data Group 0A	EON data Group 14A
1	FM-90	BF01	89.3, 90.1, 91.7	#1 B501, FM-91, 89.9, 90.7, 92.1, 92.7
				#2 BE01, NPR-88, 88.5
2	FM-91	B501	89.9, 90.7, 92.1, 92.7	#1 BE01, NPR-88, 88.5
				#2 BF01, FM-90, 89.3, 90.1, 91.7
3	NPR-88	BE01	No AF's use group 0B	#1 BF01, FM-90, 89.3, 90.1, 91.7
				#2 B501, FM-91, 89.9, 90.7, 92.1, 92.7

Figure 13-7   RDS Data for National Public Radio Network.

must always be different among stations with adjoining coverage. This will take some coordination among member stations to properly assign PI codes. To the listener, one preset recall on the receiver will always tune immediately to the strongest NPR affiliate station. This offers a great advantage to NPR as an organization, while giving all member stations a loose affiliation.

Syndicated Programs

The regional variant feature can also be utilized for syndicated programming in order to link multiple stations together during periods where the same program is broadcast. This offers the listener the advantage of remaining tuned to the program as they travel without having to retune the receiver. The operation would be similar to that described for NPR, except all stations would transmit the same nationally allocated PI code during the syndicated program, then transmit a regional variant when local programming is resumed. This process works as long as there is overlapping coverage areas between stations. The receiver preset becomes a preset for the syndicated program rather than for any particular station.

The disadvantage is that the receiver will always tune to the strongest station in this network when the preset is recalled, even when the syndicated program is not being broadcast. AF switching between stations will only occur when the syndicated program is being aired. From the point of view of pure engineering, the receiver is offering the best quality signal to the listener. From a marketing point of view, this operation is not a problem except in areas of high signal overlap between stations.

In this example, we will use only single-transmitter stations. Four stations simulcast a syndicated morning program from 5:00 to 10:00 A.M. daily.

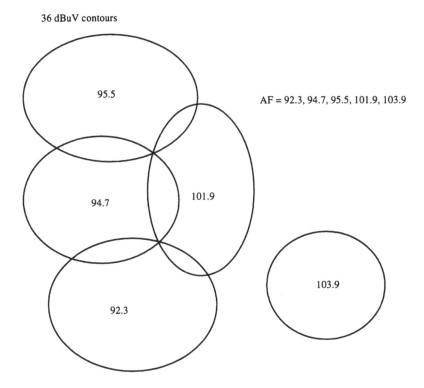

Figure 13–8 Syndicated program network.

This network is depicted in Figure 13.8. Each station transmits an AF list containing the frequencies of all five stations. Multiple transmitter stations may also be utilized simply by adding the frequencies to the AF list. Stations outside the overlapping coverage areas (i.e., 103.9) may also be added to the network AF list. While AF switching will not occur, this offers an advantage to the listener in that these stations will be tuned automatically if the listener travels within that station's coverage area and recalls the network preset.

Figure 13.9A shows the syndicated program network's RDS data. A nationally allocated PI code must be obtained for the network's use. Note that each station maintains its own identity through the use of unique Program Service (PS) names. During the airing of the syndicated program material, each station identifies the program through the use of duplicate PTY and PTYN information. This notifies the listener with visual clues that, even though the receiver may have switched to another station, they are still listening to the desired program.

Time	f	PS	PTY	PTYN	PI	Program
5 am to	95.5	The Edge	Personality	Tom&Bob	B3AB	Syndicated morning program
	101.9	Rock 102				"Tom & Bob."
	94.7	Q-95				
	103.9	Wheels				
10 am	92.3	The Fort				AF switching occurs.
10 am to	95.5	The Edge	Rock	Altrntve	B4AB	Non-syndicated, individual programming.
	101.9	Rock 102	Classic Rock	None	B5AB	No AF switching occurs if
	94.7	Q-95	Rock	None	B6AB	listener selects
5 am	103.9	Wheels	Soft Rock	None	B7AB	"reject variants"
	92.3	The Fort	Classic Rock	"70's"	B8AB	AF option.

Figure 13–9A RDS data for a syndicated radio network using the regional variant feature.

Case 4: Linking Stations with Differing PI Codes

Syndicated programs are often carried by stations with no affiliation, as marked by the differing PI codes. These stations broadcast the same audio program for periods of time. A method other than the use of regional variants offers the best method of linking stations together for a period of time, while retaining total individuality at all other times. Linkage information allows stations with differing PI codes to link and de-link at different times of the day. Linkage information consists of the following codes:

Time	f	PS	LSN	L A	PI	Program
5 am	95.5	The Edge	BA2	1	7634	Syndicated morning
to	101.9	Rock 102	BA2	1	6AC9	program
	94.7	Q-95	BA2	1	4F56	"Tom & Bob."
10 am	103.9	Wheels	BA2	1	52DD	
	92.3	The Fort	BA2	1	1F34	AF switching occurs.
10 am	95.5	The Edge	BA2	0	7634	Non-syndicated,
to	101.9	Rock 102	BA2	0	6AC9	individual programming.
	94.7	Q-95	BA2	0	4F56	
5 am	103.9	Wheels	BA2	0	52DD	
	92.3	The Fort	BA2	0	1F34	No AF switching occurs.

Figure 13-9B RDS data for a syndicated radio network using the regional variant feature.

- *Linkage Set Number (LSN)*. This 12-bit code can be assigned to a syndicated program for use. It is important that the LSN be unique to each program. The LSN is transmitted along with the PI code of the other network that may carry the simulcast program. The LSN is carried in the Enhanced Other Networks information: group type 14A, variant 12, blocks 3 and 4.
- *Linkage Actuator (LA)*. This 1-bit code is set high when the programs (as identified by the current LSN and PI of the other network) are linked, and low when they are de-linked. This information is carried in group 1A and should be transmitted at least once every five seconds.

Consumer receivers will group multiple data records under the LSN code. A data record contains all the information about a particular station or network, such as PI, PS, and AF. When the LA is set for that particular LSN, then the receiver will AF switch between all the available data records. When the LA is not set, the receiver will no longer switch between the associated data records. Using the syndicated network example shown in Figure 13.8,

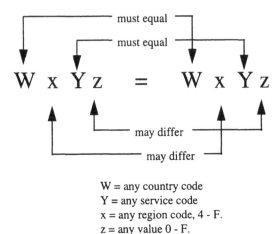

W = any country code
Y = any service code
x = any region code, 4 - F.
z = any value 0 - F.

Figure 13-10 Extended generic linkage of Program Identification codes.

along with the linkage information feature, we can derive the network configuration shown in Figure 13.9B.

Linkage offers an advantage over the use of regional PI coding in that the receiver will not tune to any other station than the one set in the preset, unless the LSNs are equal and the LA is set. The use of regional variants allows the receiver to tune to any of the stations of the network at any time when a receiver preset was recalled. The receiver can also AF switch between nonsimulcast network stations if the listener has selected the accept variants option on the receiver. Linkage information allows broadcasters to link and de-link to a variety of other stations at different times of the day without listener intervention or knowledge. The end effect is that the listener remains tuned to the simulcast or syndicated program. This results in greater listenership for the program. Each station may keep its own identity through individual PS names, and probably an even number of listeners are traded back and fourth between stations.

While this method of linking stations with differing PIs may seem an ideal solution for North America, it has been implemented on only a small percentage of receivers to date. Many RDS features have been supported in consumer products simply because broadcasters began using them. So don't let the limited application deter your usage of it. If you begin to use this or any other feature that is seldom seen in consumer equipment, let the NAB or EIA know about your application and they will get the word out to the proper people in the industry. Set aside a special place in your Web site for RDS to let everyone know what you are doing.

Extended Generic Linkage (EGL)

Extended generic linkage is an expansion of the use of regional variants that allows similar PI codes to be considered as identical. It should be noted that the use of EGL does not guarantee that the audio program is common. This can be used where common ownership of several different networks exists. Figure 13.10 shows the rule for extended generic linkage. By applying EGL, the PI code A514 can be linked to the PI code AD1A. EGL can only be utilized in North America for PI codes in the range of B_01 to B_FF; D_01 to D_FF; and E_01 to E_FF. This is the only range of PI codes that are defined for use of generic linking or regional variants.

Coding of Alternate Frequencies

There are two distinct methods of coding Alternate Frequency (AF) information into the data stream. AF information is contained in group type 0A for the tuned network and in group 14A for the Enhanced Other Network (EON) feature. The actual AF information may be transmitted in two distinct ways:

1. *Method A.* All possible AFs are transmitted in any order. Lists of up to twenty-five frequencies may be supported. The example below shows a list of five VHF frequencies. Here, #5 means that the total number of alternate frequencies is five, and is represented by code 229.

1st 0A:	#5	AF1
2nd 0A:	AF2	AF3
3rd 0A:	AF4	AF5

2. *Method B.* AFs are transmitted in distinct pairs, such that the transmitters are paired with all possible transmitters that have overlapping coverage areas. Regional variant transmitters may also be identified through a special coding scheme. Broadcasters that split a network during certain hours of the day should use AF method B, not AF method A. Lists longer than twenty-five may also be accommodated through this method.

As in AF method A, each list starts with a code giving the total number of frequencies within the list, followed by the tuning frequency for which the

F_1	F_2	Commentary
# 9	99.5	Total number (9) of frequencies for tuning frequency (99.5)
89.3	99.5	$F_2 > F_1$ hence 89.3 is an AF of tuned frequency 99.5, and is the same program
99.5	100.9	$F_2 > F_1$ hence 100.9 is an AF of tuned frequency 99.5, and is the same program
104.8	99.5	$F_2 < F_1$ hence 104.8 is an AF of a regional variant of tuned frequency 99.5
99.5	89.1	$F_2 < F_1$ hence 89.1 is an AF of a regional variant of tuned frequency 99.5

Figure 13-11   Example of AF method B coding.

list is valid. All remaining pairs give the tuning frequency together with a valid AF. For the transmission of the frequency pairs, the following convention is used:

- They are generally transmitted in ascending order, e.g.

| 89.3 | 99.5 | or | 99.5 | 101.8 | $F_1 < F_2$

- Where regional variants are found, the frequencies are transmitted in descending order, e.g.

| 99.5 | 90.6 | or | 100.7 | 99.5 | $F_1 > F_2$

In both of the examples above, 99.5 MHz is the tuning frequency. An AF method B list may look like the one shown in the example of Figure 13.11 for the transmitter network shown in Figure 13.12.

AF Code Tables and Special Meanings

The actual AF information is coded into the data groups. Figure 13.13 shows the coding structure for the FM frequencies. Special codes are also utilized

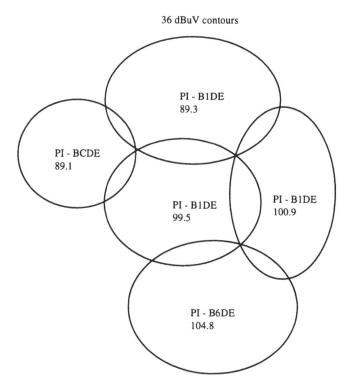

36 dBuV contours

Figure 13-12 Transmitter network for AF method B example.

Number	Binary code	Carrier frequency
0	0000 0000	Not to be used
1	0000 0001	87.6 MHz
2	0000 0010	87.7 MHz
:	:	:
:	:	:
204	1100 1100	107.9 MHz

Figure 13-13
Alternate Frequency
VHF code table.

to provide the necessary information about the AF list, such as the number of AFs in the list, or a filler code when the actual data does not fill out the entire group. These special codes are given in Figure 13.14. AF codes in type

Number	Binary code	Special meaning
0	0000 0000	Not to be used
205	1100 1101	Filler code
206	1100 1110	Not assigned
:	:	:
223	1101 1111	Not assigned
224	1110 0000	No AF exists
225	1110 0001	1 AF follows
:	:	:
249	1111 1001	25 AF's follow
250	1111 1010	An LF/MF frequency follows
251	1111 1011	Not assigned
:	:	:
255	1111 1111	Not assigned

Figure 13-14
Alternate Frequency
special meanings
code table.

14A groups are used to refer to frequencies of other networks. Both method A and B are supported in this structure. Variant 4 utilizes AF method A coding to transmit up to twenty-five frequencies. The PI code of the other network to which the AF list applies is given in block 4 of the group. Variant 5 is used for the transmission of "Mapped frequency pairs." This is used specifically to reference a frequency in the tuned network to a corresponding frequency in another network. This is particularly useful for a broadcaster who transmits several different services from the same transmitter tower with the same coverage areas. The first AF code in block 3 refers to the frequency of the tuned network; the second code is the corresponding frequency of the other network identified by the PI code in block 4. See Chapter 14 on Enhanced Other Networks for more details of this feature.

14

Enhanced Other Networks (EON)

The EON function allows data to be shared among affiliated networks. This affiliation may be joint ownership, or simply an agreement to share information about other stations in the area. All stations are cross-referenced by means of their Program Identification (PI) codes. Features that may be transmitted using EON for other program services are: Alternate Frequencies (AF), Program Item Number (PIN), Program Service Name (PS), Program Type code (PTY), Traffic Announcement (TA), Traffic Program (TP), and Linkage. This information is carried in type 14 groups. These features are covered in their respective chapters of this book, so a detailed description of each feature will not be given here.

EON features are utilized by EON-equipped consumer receivers in the following ways:

1. *Automatic updating of receiver memory.* Receivers can update the memory content to store information about other networks or stations that it is not currently tuned to. The PI, PS, and list of AFs can be given for referenced networks. This allows the receiver to tune instantly to an associated network without having to initiate a frequency search. Once tuned to an EON broadcast, the PS name can be displayed as soon as the PI code is received. Additionally, PTY searches are instantaneous, since the frequency that the selected program is on is already known.

2. *Receive service broadcasts from other networks.* Traffic announcements on affiliated networks can be tuned automatically through EON.

This allows localized traffic information to be given without causing interruption to the main network. Receivers incorporating the PTY watch or interrupt feature can use the EON PTY information to switch to other networks for the selected program. Similarly, receivers equipped with the PIN feature can also respond to programs referenced via EON.

3. *Linkage information.* Linkage information provides the means by which several Program Services, each characterized by its own PI code, may be treated by a receiver as a single service during times in which a common program is carried.

The type 14 group has two versions, A and B. The A version is the normal form and is used for the background transmission of Enhanced Other Networks information. The maximum cycle time for the transmission of all data relating to all cross-referenced Program Services shall be less than two minutes. The A version has sixteen variants that may be used in any mixture and order. Both AF method A and AF method B exist for the transmission of frequencies of cross-referenced Program Services. The most appropriate AF method should be chosen for each cross-referenced Program Service.

The B version of a type 14 group is used to indicate a change in the status of the TA flag of a cross-referenced Program Service and should only be transmitted during a change in the TA status of a cross-referenced service. If a transmitter cross-references to more than one traffic program with a different PI code via the EON feature, the time between two different references via the type 14B group has to be longer than two seconds. Some early RDS EON consumer receivers may need up to four correct type 14B groups for reliable functioning. Therefore, it is recommended to broadcast as many as possible of up to eight type 14B groups, in order to ensure the detection of the switching under bad reception conditions.

At the time of development of the EON feature, no provision was made for automatic retuning for PTY or PIN information such as exists for EON TA. Since there is no dedicated group such as 14A for the PTY or PIN information, it is recommended that multiple bursts of the appropriate 14A group variant be given whenever the PTY or PIN information for a cross-referenced network changes. Bursts of up to eight are advised to ensure proper reception even under adverse reception conditions.

While the use of EON is not straight forward for the U.S. broadcast infrastructure, there are many advantages to its use. For instance, a station that does not normally carry traffic information can reference an affiliated station so that the listeners may obtain the selected service without having to tune away from the current station. This allows the best of both worlds for the listener, and may even allow an overall cost savings for the broadcaster if the traffic is purchased from a secondary source.

15

Other RDS Features

This chapter includes many important features of RDS that are either very simple or not covered in explicit detail within the scope of this book. The more complex issues will include references for more detailed descriptions in other sources.

Music/Speech (M/S) Switch Code

This is a 1-bit code. A "0" indicates that speech is being broadcast at present, and a "1" indicates that music is being broadcast at present. When the broadcaster is not using this facility, the bit value will be set at "1." Some consumer receivers incorporate this feature to allow control of the audio-tone controls such that speech-oriented broadcasts are optimized automatically for the listener. One concept is to allow the listener to set the speech-tone preference. The receiver's stereo/mono switch could also be set to mono during speech reception to decrease audio noise. The M/S code is carried in group types 0A, 0B, and 15B.

Decoder Identification and Dynamic PTY Indicator/DI Codes

The decoder information is contained in four bits that are used to indicate different operating modes to switch individual decoders on or off, and to indicate if PTY codes in the transmission are dynamically switched. The decoder information feature was modified in the revised *Standard*, adding a

Settings	Meaning
Bit d_0, set to 0:	Mono
Bit d_0, set to 1:	Stereo
Bit d_1, set to 0:	Not Artificial Head
Bit d_1, set to 1:	Artificial Head
Bit d_2, set to 0:	Not compressed
Bit d_2, set to 1:	Compressed
Bit d_3, set to 0:	Static PTY
Bit d_3, set to 1:	Indicates that PTY code on tuned service or referenced service in EON variant 13 is dynamically switched

Figure 15-1
Decoder information codes.

new feature called the dynamic PTY indicator. Many consumer receivers include a feature that allows the listener to standby or watch for a favorite PTY to become available. For instance, a listener may want to be interrupted when the News or Weather is broadcast. The dynamic PTY indicator allows the broadcaster to signal to the receiver that the PTY codes are dynamically switched as the program content changes. One possible use of this information is to inform the listener that the PTY standby mode is unavailable on the currently tuned station. The DI codes are carried in group types 0A, 0B, and 15B. Figure 15.1 shows the possible states and meanings for the DI codes.

Program Item Number (PIN) Codes

The PIN feature is utilized where scheduled, published programs are regularly broadcast. The PIN is a code of the program's scheduled date and start time that is broadcast when the program actually begins. Since programs can often be delayed by several minutes due to variations in the schedule, the PIN allows consumer receivers to be signaled when the program begins. Some consumer receivers include a PIN feature that allows the listener to select programs to be recorded automatically. When the desired program begins, as signaled by the PIN transmission, the consumer receiver will record the selected program automatically, allowing later playback. The PIN codes are carried in group type 1A and 1B. The transmitted Program Item Number code will be the scheduled broadcast start time and day of month, as published by the broadcaster. If a type 1 group is transmitted without a valid PIN, the day of the month shall be set to zero. In this case, a receiver that evaluates PIN shall ignore the other information in block 4.

Language Identification (LI)

The Language Identification code allows the broadcaster to identify the spoken language that is currently being transmitted. This is especially useful when the language is the non-native language of the country. Many ethnic programs are carried in various portions of the country. By utilizing the Language Identification code, it is possible for consumer receivers to search for stations that carry the preferred language of the listener. The Language Identification code is carried in group 1A variant 3. It should be transmitted at least once every two seconds when being used. This code is detailed in Annex J of the *Standard.*

RDS and the Emergency Alert System (EAS)

Annex R of the *RBDS Standard,* entitled "Emergency Alert System Open Data Application," makes use of the ODA protocol to provide RDS solutions to implementing the EAS system of the FCC. The EAS system affects only stations in the U.S. and its territories. A detailed description of the EAS protocol is given in 47 CFR Section 11 of the *Federal Communications Commission (FCC) Rules and Regulation.* The RDS EAS system has been designed to augment rather than replace existing RDS features related to emergency warning, providing a silent means of carrying SAME encoded data, as well as providing timely warnings and information to your listeners. The FCC recognizes the voluntary use of the *RBDS Standard* within its final rules in order to provide a useful structure to the EAS. Within the *RBDS Standard* there are several other methods for carrying emergency information besides the EAS ODA. Only one system may be chosen for use. These alternate warning systems are:

1. *Group 1A address code 7/Group 9A.* The reception of this group type indicates that the broadcaster carries EAS transmissions. Emergency announcements will be signaled by PTY 31. Other RDS features, such as radiotext, may also provide emergency information. There are no published data protocols for the use of the 9A group in the U.S. This group type combination is utilized by SAGE Alert corporation.
2. *Via the MBS/MMBS system.* See Annex P of the *RBDS Standard* for details.

As mentioned above, the EAS ODA makes use of existing RDS features while adding vitally needed information to meet FCC and user requirements. The following existing RDS features are utilized with the open protocol:

1. *PTY-30 "Test."* Indicates that a test transmission is in process. No interruption of the audio should occur during the reception of this code. Received data will be handled as test data and not valid warning data.

2. *PTY-31 "Alert."* Indicates that an emergency alert is in process. An audio warning message will accompany this code. Consumer equipment should interrupt current operations (i.e., playback or radio off) during the reception of this code and switch over to FM reception. Receivers should increase the volume to an audible level during the reception of this code.

3. *Radiotext (Group type 2A/B).* Broadcast EAS equipment should decode and reconstruct EAS messages into this format for reception by consumer receivers. This prevents the necessity for having SAME data conversion software in each consumer receiver.

As can be seen, the EAS ODA is best utilized when melded with existing features. This makes the actual implementation easier for consumer and warning receivers. The following new features were added through the EAS ODA to provide vital information:

1. *The identification of an EAS broadcast station.* Reception of the proper AID code, E911, indicates that the currently tuned station provides EAS data in accordance with this protocol. This data is transmitted at a minimum of once per second, thus allowing automated search tuning.

2. *FCC EAS Compatible.* The EAS open protocol includes the retransmission of all SAME data. This data can serve as a secondary link in the EAS service "web." In this manner, EAS data can be carried silently by FM broadcasters.

3. *Error reduction of SAME data.* The SAME data may be transmitted multiple times, allowing the use of time diversity to ensure accurate message delivery. In fact, the SAME data may be transmitted constantly for the duration of the event. Traditional RDS error detection and correction may also be employed to ensure data integrity.

4. *Provision for private or encrypted Emergency Services.* Companies who desire to carry encrypted emergency data may do so by applying for a System Identification code through the NRSC. Spare data fields contained in the 3A and 9A groups may be utilized to carry this data. Transmission of SAME data must be given priority within the system. This feature allows localized or even national fee-based emergency data services to be supported.

5. *Operation of a sleep/wake cycle for battery-powered equipment.* The Warning Activation (WA) bit will be set prior to the transmission of any test or alert data, allowing receivers to "wake" upon the reception of WA=1. In this manner a receiver may "sleep" for 9 seconds, and "wake" long enough to receive the WA bit, thus greatly conserving battery power.

6. *Identification of alternate EAS providers.* Secondary EAS providers may be identified through this feature for storage in the receiver. This allows instant tuning to an alternate frequency should the currently tuned station go off the air during an alert message. This feature also allows automatic tracking of EAS stations by a mobile receiver. The EAS Other Network (ON) data shall be kept separate from other AF information, since the station's audio broadcast will not be coordinated except during an actual Alert situation.

7. *Instant tuning to alternate EAS broadcasts.* The broadcaster can automatically retune the listener to an alternate network that is carrying Alert information. The broadcasts must be coordinated such that the PTY of the other network is set to 31 within two seconds after the switching data is transmitted.

For a further description of the EAS ODA protocol, see Annex R of the *RBDS Standard.*

AM RDS

At this time, an RDS-like data system has not been developed for AM broadcast use. An AM data system was developed in Europe, but it is not compatible with the C-QUAM AM stereo system utilized in the U.S. At least one company is performing development work on an AM data system that works in the adjacent channels bandwidth, but under the FCC's required masks of adjacent channel interference. The *RBDS Standard* includes provision for the addition of an AM data system should one be developed.

ID Logic and IRDS Updating

A separately licensed technology called ID logic is available for the use of receiver manufacturers to provide RDS-like information about AM and non-RDS FM stations. The ID logic system utilizes an In-receiver Database (ID) that contains the call letters, frequency, and format type for all AM and FM stations. The receiver determines the geographic area of use, with minimal help from the user, and can then identify the tuned station. The call letters of the station can then be displayed on the receiver display, somewhat analogous to the RDS Program Service (PS) name display. Additionally, the format of the station can be identified. The user can also search for a desired format in a manner similar to that of the RDS program type (PTY) search. The formats used in ID logic do not match the defined RDS PTY codes but can be constructed into a single list. Therefore, it is possible to construct a consumer receiver that provides several of the RDS features, even on AM and non-RDS FM stations. Due to the fact that station information—such as frequency, call letters, and format—can change, or new stations may be added, some provision for updating the receiver database must be provided. The *RBDS Standard* includes an update protocol utilizing the Open Data Application (ODA) to update the receiver database automatically. While ID logic information may not be as dynamic as RDS, it does offer the listener the appearance of RDS for all stations. ID logic is licensed by the PRS corporation.

16

Open Data Channel

Use of Open Data Applications

The open data applications (ODA) is a recent development for the *RDS Standard*. This powerful new application was developed after some frustrating attempts to develop new RDS features. The problem was that very few RDS data groups were remaining that could be defined for a new application. Once these groups were defined, there were no more future applications that could be developed. Achieving a worldwide consensus for the precious commodity of RDS data groups seemed impossible until the idea for ODA was developed. The concepts are simple:

- Allow the user to design a custom data group for a specific application. The group must retain the basic structure of data structure, but aside from that, the user defines the usage.
- The user must have some means of detecting his application. The Application Identification (AID) code was developed so that receivers can search for and positively identify their data broadcast.
- The application should be flexible enough to be moved to various data groups without causing the receiver to become confused. If this is done, then nonprimary data groups can also be "reused" as an ODA.

Application Identification for Open Data

Figure 16.1 shows the Open Data Application group type 3A. This group provides the following information:

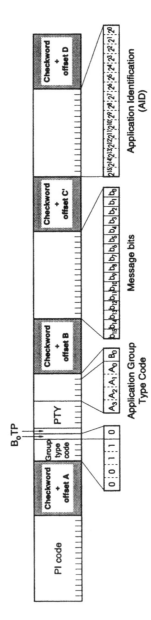

Figure 16–1 Application Identification for Open data, Type 3A group.

1. *The Application Identification (AID) Code.* The four-digit hexadecimal number is assigned by either the NAB or EBU for use by an applicant on an international basis. Data receivers search for this code to identify the proper service.
2. *Application group type.* This data field indicates the group where the application data is located on the particular transmitter. Receivers must be designed to decode the application data on the referenced group. Special codes are assigned to indicate an encoder data fault, or when all the data for the application is contained with the 3A group.
3. *Data field.* Up to sixteen data bits may be used for the application. Small applications may be self-contained within the 3A group. This field could also be used for subaddressing for subscriber-based services.

Application Data Groups

Two basic ODA data group types can be readily identified:

1. Groups only assigned for ODA.
2. Groups with a dual function. These groups are assigned as a secondary RDS feature. When not used for that feature, they may be used for ODA. These groups can be further divided as:

- Groups that are signaled through the slow-labeling group, 1A, such as radio paging (group 7A) and traffic message channel (group 8A).
- Those that are stand-alone secondary features, such as transparent data channel (group 5A/B), and in-house data channel (group 6A/B).

In effect, the development of the ODA has given the *RDS Standard* a real breath of life. Datacasting services may now be started very easily without fear of crashing with some other system. The AID code will serve as the "datacasting coordinator," allowing services to be operated independent of the PI code. We will first explore the structure of the ODA, then take a close look at how to obtain an AID code and begin your data service.

An ODA may use type A and/or type B groups, however it must not be designed to operate with a specific group type. The specific group type used by the ODA in any particular transmission is signaled in the Application Group Type Code carried in type 3A groups. Figure 16.2 shows the type A and type B groups that may be allocated to ODA. Group types not shown in this figure are not available for ODA. Note that data groups devoted to consumer-receiver features may not be allocated for use as an ODA.

Group type	Application group type code	Availability for Open Data Applications
	00000	Special meaning: Data is self contained within group 3A
3B	00111	Available unconditionally
4B	01001	Available unconditionally
5A	01010	Available when not used for TDC
5B	01011	Available when not used for TDC
6A	01100	Available when not used for IH
6B	01101	Available when not used for IH
7A	01110	Available when not used for RP
7B	01111	Available unconditionally
8A	10000	Available when not used for TMC
8B	10001	Available unconditionally
9A	10010	Available when not used for EWS
9B	10011	Available unconditionally
10B	10101	Available unconditionally
11A	10110	Available unconditionally
11B	10111	Available unconditionally
12A	11000	Available unconditionally
12B	11001	Available unconditionally
13A	11010	Available when not used for RP
13B	11011	Available unconditionally
	11111	Special meaning: Temporary data fault (Encoder status)

Figure 16–2 Group types available for Open Data Applications (ODA).

Open Data Applications: Group Structure

Open Data Applications must use the format shown in Figure 16.3 for ODA type A groups, and the format in Figure 16.4 for ODA type B groups. Note that type B groups always carry the PI code in block 3, even if that group is only utilized within an ODA. This is done to ensure compatibility with consumer receivers. Receivers will always recognize the third block as the PI code for type B groups regardless of the actual group number. To ensure backwards compatibility even an ODA must follow this structure.

Choosing the Proper Data Group for Your Application

When developing an ODA, you must choose how your data will be transmitted. Figure 16.5 summarizes the available data bits for an ODA. Small applications can exist totally within the 3A group. If this applies to your application, then the special group application code of "00000" should be utilized. If your application requires more than sixteen bits, then you must reference another group type for your application. While the ODA is designed to be flexible in construction, your application will probably only exist within an "A" or a "B" type data group. Your application must, however, be designed to accommodate any "A" or "B" type group. This allows the data receiver to be flexible in operation, because each broadcaster may not have the same data group open for an ODA application.

The ODA Directory specification associated with a particular AID code defines the use of type A and type B groups as various modes of operation. It is possible to operate an Open Data Application using both A and B type groups together. In this mode, the maximum number of data bits is obtained. These must reside within the same group number, since there is no way to signal the receiver to any other mode of operation. The possible modes of operation are:

- *Mode 1.1.* Type A groups used alone.
- *Mode 1.2.* Type B groups used alone.
- *Mode 2.* Type A groups and type B groups used as alternatives.
- *Mode 3.* Type A groups and type B groups used together. Applications that actively utilize both type A and B groups are signaled using two type 3A groups.

The receiver software must know the mode of operation, but does not require that the data be located in a fixed group number. The group number is assigned at the transmission site according to availability. Mode 3 and 4 can only be utilized with data groups that have both A and B types available for the ODA.

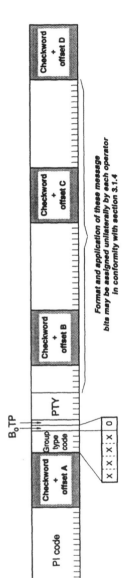

Figure 16-3 ODA type A groups.

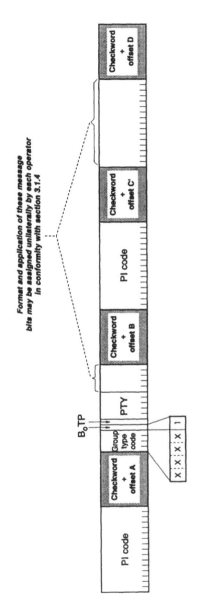

Figure 16-4 ODA type B groups.

Group	Data bits available
3A (Required)	16
Any "A" type	37
Any "B" type	21

Figure 16-5
Data bits available for
Open Data Applications.

Locating Your ODA

The data receiver must locate and synchronize to the appropriate ODA. The process is as follows:

1. The receiver seeks to the next receivable station, receivable being defined as one of sufficient signal strength to provide an acceptable bit-error rate.
2. The receiver waits to receive group type 3A. If after a sufficient amount of time, the receiver detects no RDS data, then the receiver should seek to the next station.
3. If RDS is detected, then the receiver waits a sufficient amount of time for reception group 3A.
4. If group 3A is not detected, then the receiver should seek to the next station.
5. If group 3A is detected, then the received AID code must be verified against the desired AID code.
6. If the received AID code does not match the desired AID code, then the receiver should seek to the next station.
7. Once the AID code is verified, the receiver should check the status of the application group type code. This step applies only to applications referenced outside the 3A group (i.e., group application code not equal to "00000").
8. The received application group type code is next stored to guide the receiver to the location for the remaining data.
9. The application software should check the status of the application group type code for code "11111," which indicates that a temporary data fault has occurred. This implies that the received data should be ignored until the condition is cleared.

Multiple Applications

It is possible that multiple ODAs can reside on a single transmitter or network. Obviously, only a fixed amount of data may reside within the constraints of the RDS system. It is the transmission operator who must trade the benefits of increased income versus the needs of the station. Some ideas that may be employed to increase system throughput are:

- Utilize the most efficient data coding possible.
- If a particular application requires use only during a portion of the day, then consider transmitting that application during that time only. The unused time can be resold or utilized for other purposes.
- Dynamic increases of changed data can insure proper reception of the data without adversely impacting the overall throughput.

Applying for an Application Identification Code

An AID code may be obtained by filling out an application form that is contained in the *Standard*. A nominal registration fee is collected by the issuing agency. AID codes may be obtained through the European Broadcasting Union (EBU) or the National Association of Broadcasters (NAB). The codes issued by either organization are internationally allocated but are separately issued for the convenience of the applicant. Once a code is issued, it may be used by the applicant throughout the world, allowing the possibility of "growing" a data service. Over 65,000 AID codes are available for use.

Data Receivers

A number of companies offer RDS-data receivers. The data receivers can interface with switched outputs or RS-232 interfaces, for example. These receivers must have application-specific software written. Often, the same companies offering the receivers can also offer software development. Data receivers are currently offered by Advanced Digital Systems (PC-based), Modulation Sciences, Nokia, and Aztech. Consider the variety of requirements that your application requires when selecting a data receiver.

17
RDS Group Structure

After the RDS signal is modulated and demodulated as a synchronous bit stream, this bit stream or endless stream of logical 1's and 0's are structured into a baseband code. The RDS group structure or baseband coding is depicted in Figure 17.1. This baseband structure is described as follows:

- The largest element in the structure is called a "group" of 104 bits each.
- Each group comprises four blocks of 26 bits each.
- Each block comprises an information word and a checkword.
- Each information word comprises 16 bits.
- Each checkword comprises 10 bits.
- All information words, checkwords, binary numbers, or binary address values have their most significant bit (m.s.b.) transmitted first.

The data transmission is fully synchronous, and there are no gaps between the groups or blocks.

Information Word

The information word contains the actual data for each block. This information word is comprised of sixteen bits. The most significant bit (m^{15}) is transmitted first.

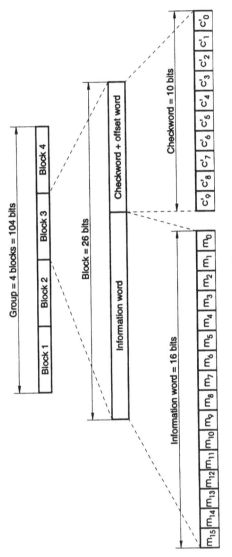

Figure 17-1   Structure of the baseband coding.

Checkword and Offset Word

The checkword and offset consist of ten bits. The most significant bit (c'_9) is transmitted first. The checkword allows the receiver to detect and correct errors that occur during data reception. The error-protecting code has the following error-checking capabilities:

- Detects all single- and double-bit errors in a block.
- Detects any single-error burst spanning 10 bits or less.
- Detects about 99.8% of bursts spanning 11 bits and about 99.9% of all longer bursts.
- The code is also an optimal burst error-correcting code and is capable of correcting any single burst of a span of 5 bits or less.

Figure 17.2 shows the possible offset words. The offset word allows detection of the block number within the data group. This allows the PI, PTY, and TP codes to be decoded without reference to any block outside the one that contains the information. This is essential to minimize acquisition time for these kinds of message and to retain the advantages of the short (26-bit) block length. To permit this to be done for the PI codes in block 3 of version

Offset word	Binary value									
	d_9	d_8	d_7	d_6	d_5	d_4	d_3	d_2	d_1	d_0
A	0	0	1	1	1	1	1	1	0	0
B	0	1	1	0	0	1	1	0	0	0
C	0	1	0	1	1	0	1	0	0	0
C'	1	1	0	1	0	1	0	0	0	0
D	0	1	1	0	1	1	0	1	0	0
E	0	0	0	0	0	0	0	0	0	0

Figure 17–2 Offset words.

B groups, a special offset word (C') is used in block 3 of version B groups. The occurrence of offset C' in block 3 of any group can then be used to indicate directly that block 3 is a PI code, without any reference to the value of B_0 in block 2. The offset word E is only transmitted by MBS or MMBS stations. This allows receivers to maintain synchronization in MMBS applications.

Synchronization of Blocks and Groups

The blocks within each group are identified by the offset words A, B, C or C', and D added to blocks 1, 2, 3, and 4, respectively, in each group. The beginnings and ends of the data blocks may be recognized in the receiver decoder by using the fact that the error-checking decoder will, with a high level of confidence, detect block synchronization slip as well as additive errors. This system of block synchronization is made reliable by the addition of the offset words, which also serve to identify the blocks within the group. These offset words destroy the cyclic property of the basic code so that, in the modified code, cyclic shifts of codewords do not give rise to other codewords. A detailed explanation of a technique for extracting the block synchronization information at the receiver is given in Annex C of the *Standard*.

Group Structure

The information coded into each group has a common fixed structure, as depicted in Figure 17.3.

All data groups share a common fixed structure of information coding to allow critical data to be transmitted in a fixed, highly repetitive pattern to ensure the best reception, even under adverse reception conditions. This fixed structure is described as follows:

1. *Block 1.* Block 1, or every RDS data group, contains only the program identification or PI code.
   a) Block 1 is identified by offset word "A."
2. *Block 2.* Block 2 contains the group type code, version code, traffic program code, and program type (PTY) code. All codes are binary coded with the m.s.b. transmitted first.
   a) *Group type code.* The data groups are identified through the 4-bit group type code identified as A3 to A0, along with the 1-bit version code B0. The bits A3 to A0 yield decimal values 0 through 15.
   b) *Version code.* Each group then has an "A" version and a "B" version based upon the state of B0 where:

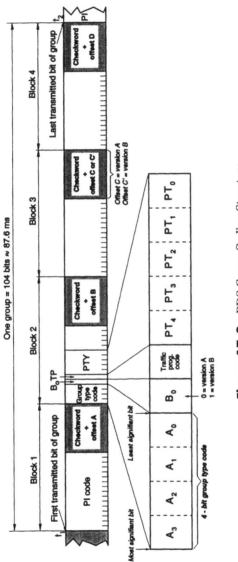

Figure 17-3 RDS Group Coding Structure.

      i) B0=0 is identified as an "A" group type.

      ii) B0=1 is identified as a "B" group type.

      iii) All "B" type groups repeat the program identification (PI) code in block 3. All "B" group types carry the PI code in the third block as well as the first.

    c) *Traffic Program (TP) code.* This single bit code is utilized with the Traffic Announcement (TA) code to provide the traffic feature.

    d) *Program Type (PTY) code.* This 5-bit code describes the current audio program being aired. Thirty-one possible PTYs are available to choose from.

    e) Block 2 is identified by offset word "B."

  3. *Block 3.* Dependent upon the group version code (A or B) or offset word (C or C'), block 3 may carry two distinct types of data:

    a) *Version code "A."* This block carries data defined by the group type code.

      i) Block 3 is identified by offset word "C" for version code "A" groups.

    b) Version "B" type groups carry only the PI code.

      i) Block 3 is identified by offset word C' for version code "B" groups.

  4. *Block 4.* This block carries data defined by the group type code.

    a) Block 4 is identified by offset word "D."

RDS Group Types

RDS data groups are referred to by both the group type and version, that is, 0A and 14B. Taking both the group type and version codes together, it can be stated that there are thirty-two defined data groups. Some of the group features are defined in their entirety and cannot be defined in any other way, while other data groups are openly defined, such that the actual data contained in the group can be defined by the operator. The group type code defines the basic feature function of the data group. The "A" and "B" versions differ only in the coding of the data for the group, as RDS data is divided into functional groups that share a common structure but perform different functions. The operator can select from these group types and transmit all or only a portion of them, depending upon the features in use. In this manner, there is no need to transmit data that is not in use, thereby maximizing data throughput. Figure 17.4 summarizes the possible applications for all possible group types. Because some groups' usage is also tied to the use of the slow-labeling group (1A), it is included for reference. These groups may be used for Open Data Applications when not flagged for use in group type 1A.

Group type	Group type code/version					Flagged in type	Description
type	A₃	A₂	A₁	A₀	B₀	1A groups	
0 A	0	0	0	0	0		Basic tuning and switching information only
0 B	0	0	0	0	1		Basic tuning and switching information only
1 A	0	0	0	1	0		Program Item Number and slow labeling codes only
1 B	0	0	0	1	1		Program Item Number
2 A	0	0	1	0	0		Radiotext only
2 B	0	0	1	0	1		Radiotext only
3 A	0	0	1	1	0		Applications Identification for ODA only (see 3.1.5.5)
3 B	0	0	1	1	1		Open Data Applications
4 A	0	1	0	0	0		Clock-time and date only
4 B	0	1	0	0	1		Open Data Applications
5 A	0	1	0	1	0		Transparent Data Channels (32 channels) or ODA
5 B	0	1	0	1	1		Transparent Data Channels (32 channels) or ODA
6 A	0	1	1	0	0		In House applications or ODA
6 B	0	1	1	0	1		In House applications or ODA

continues

Figure 17–4 RDS Group Types.

Group Repetition Rates

Figure 17.5 summarizes the location of the primary RDS data features in the data groups, as well as the recommended or inherent repetition rates.

7 A	0	1	1	1	0	Y	Radio Paging or ODA
7 B	0	1	1	1	1		Open Data Applications
8 A	1	0	0	0	0	Y	Traffic Message Channel or ODA
8 B	1	0	0	0	1		Open Data Applications
9 A	1	0	0	1	0	Y	Emergency Warning System or ODA
9 B	1	0	0	1	1		Open Data Applications
10 A	1	0	1	0	0		Program Type Name
10 B	1	0	1	0	1		Open Data Applications
11 A	1	0	1	1	0		Open Data Applications
11 B	1	0	1	1	1		Open Data Applications
12 A	1	1	0	0	0		Open Data Applications
12 B	1	1	0	0	1		Open Data Applications
13 A	1	1	0	1	0	Y	Enhanced Radio Paging or ODA
13 B	1	1	0	1	1		Open Data Applications
14 A	1	1	1	0	0		Enhanced Other Networks information only
14 B	1	1	1	0	1		Enhanced Other Networks information only
15 A	1	1	1	1	0		Defined in RBDS only
15 B	1	1	1	1	1		Fast switching information only

Figure 17–4 *Concluded*

With a baud rate of 1187.5 bits per second, or 11.4 groups per second, it is easy to see that not every feature of RDS can be utilized at the same time. In fact, when group overhead for the checkword, offset, and group address is accounted for, the actual usable data rate is reduced to 673.7 bits per second.

Taking into account the usable data rate along with the number of occupied bits per second of each feature, the percent of total capacity can be determined. Figure 17.6 shows the data-capacity requirements for each feature. Transmission of ancillary data must be carefully calculated in com-

Main Features	Group types containing this information	Appropriate group repetition rate per sec.
Program Identification (PI) code	all	11.4
Program Type (PTY) code	all	11.4
Traffic Program (TP) identification code	all	11.4
Program Service (PS) name	0A, 0B	1
Alternative Frequency (AF) code pairs	0A	4
Traffic Announcement (TA) code	0A, 0B, 15B	4
Decoder Identification (DI) code	0A, 0B, 15B	4
Music/ Speech (M/S) code	0A, 0B, 15B	1
Radiotext (RT) message	2A, 2B	4 [1])
Program Type Name (PTYN)	10A	2 [2])
Enhanced Other Networks information (EON)	14A, 14B	0.2 up to 2 [3])

[1])   A total of 16 type 2A groups are required to transmit a 64 character radiotext message and therefore 3.2 type 2A groups will be required per second, assuming a total cycle time of 5 seconds.

[2])   This assumes a cycle time of 1 second for the entire PTYN. A slower rate may be acceptable.

[3])   The maximum cycle time for the transmission of all data relating to all cross-referenced program services shall be less than 2 minutes.

Figure 17-5   Main Feature location and repetition rates.

parison to the capacity required to support the consumer features chosen. The capacity requirements of services such as paging and Open Data Applications are limited due to the high capacity required for consumer features.

Application	% of 637.7 BPS	Accumulated capacity
PI	27.07%	
PTY	8.46%	
TP	1.69%	
PS	2.37%	39.60%
TA	0.59%	
DI	0.59%	
M/S	0.15%	
CT	0.10%	41.04%
AF	9.50%	50.54%
PTYN	10.09%	60.63%
RT	21.97%	82.60%
EON	1.10%	
PIN	0.11%	83.81%

Figure 17–6 RDS capacity requirements.

18

RDS Data Group
Coding

In this chapter we will examine all the various data groups and their application. The data groups are determined by the group type code (0 through 15)
and version code (A or B). In the previous chapter, the usage for each of
these groups was outlined. All groups that are defined only for use in the
Open Data Application will be shown in abbreviated format. The exact definition of the data bits will be defined by the service application. The intent of
this chapter is to provide the concise group bit definition and technical usage. Detailed, practical usage of the data groups are contained in their respective chapters. For ease of use, the data groups will be presented by their
usage rather than in numerical order. The groups are divided into the following headings:

1. *Consumer receiver applications.* These groups provide data about
   the broadcast and related program services intended for use by
   consumer-receiving equipment. Not all receivers will implement
   each feature.
2. *Service oriented applications.* These groups provided data that is not
   normally intended for use by consumer receivers. Applications such
   as paging and emergency warning data are included.
3. *Datacasting applications.* These groups provide a means for broadcasting a wide scope of data for use with data receivers. Typically,
   these applications are service oriented. Figure 18.1 summarizes the
   use for each group type.

Group	Description
0 A	Basic tuning and switching information only
0 B	Basic tuning and switching information only
1 A	Program Item Number and slow labeling codes
1 B	Program Item Number
2 A	Radiotext only
2 B	Radiotext only
3 A	Applications Identification for ODA only
3 B	Open Data Applications
4 A	Clock-time and date only
4 B	Open Data Applications
5 A	Transparent Data Channels (32 channels) or ODA
5 B	Transparent Data Channels (32 channels) or ODA
6 A	In House applications or ODA
6 B	In House applications or ODA
7 A	Radio Paging or ODA
7 B	Open Data Applications
8 A	Traffic Message Channel or ODA
8 B	Open Data Applications
9 A	Emergency Warning System or ODA
9 B	Open Data Applications
10 A	Program Type Name
10 B	Open Data Applications
11 A	Open Data Applications
11 B	Open Data Applications
12 A	Open Data Applications
12 B	Open Data Applications
13 A	Enhanced Radio Paging or ODA
13 B	Open Data Applications
14 A	Enhanced Other Networks information only
14 B	Enhanced Other Networks information only
15 A	Defined in RBDS only
15 B	Fast switching information only

Figure 18-1
RDS Group Types.

CONSUMER RECEIVER-ORIENTED APPLICATIONS

Type 0 Groups: Basic Tuning and Switching Information

Use this group if:

- All stations must broadcast this group at least every 500 ms.
- Use the A type group if your station has alternate frequencies.
- Use the B type group if your program is carried only on a single transmitter.

As the name implies, the type 0 group is *the* basic RDS data group. All consumer receivers decode this data group, for it contains the Program Service name and alternative frequency information. The 0A group shown in Figure 18.2 is used whenever the broadcast carries the identical program on alternate frequencies. The 0B group shown in Figure 18.3 is used when the broadcast is carried only from a single transmitter. It is essential that either of these data groups be transmitted at least once every 500 ms in order for consumer receivers to operate properly.

The Program Service (PS) name comprises eight characters that are intended for static display on a receiver. Typically, the PS is displayed instead of the actual tuning frequency, since the broadcast may be carried on multiple frequencies. The use of the PS to transmit text other than a single eight-character name is not permitted under any circumstances. Consumer receivers typically store the PS name in memory for instant display once the program identification (PI) code is detected. It is important that the PS text be static, since it is constantly displayed on the receiver and is not controllable by the listener. The radiotext groups 2A and 2B are specifically designed for dynamic text messaging. The data contained specifically within group 0 includes:

1. TA = Traffic Announcement code (1 bit)
2. M/S = Music/Speech switch code (1 bit)
3. DI = Decoder-Identification control code (4 bits). This code is transmitted as 1 bit in each type 0 group. The Program Service name and DI segment address code (C_1 and C_0) serves to locate these bits in the DI codeword. Thus, in a group with C_1C_0 = "00" the DI bit in that group is d_3. These code bits are transmitted most significant bit (d_3) first.
4. Program Service name (for display) is transmitted as eight-bit characters. Eight characters (including spaces) are allowed for each network or station and are transmitted as a two-character segment in each type 0 group. These segments are located in the displayed name by the code bits C_1 and C_0 in block 2. The addresses of the characters increase from left to right in the display. The most significant bit (b_7) of each character is transmitted first.
5. Alternative frequency codes (2×8 bits). There are two methods (A and B) for transmission of alternative frequencies.

Type 1 Groups: Program Item Number and Slow-Labeling Codes

Use this group if you utilize or have implemented any of the following features:

- Extended Country codes
- Program Item Number (PIN)

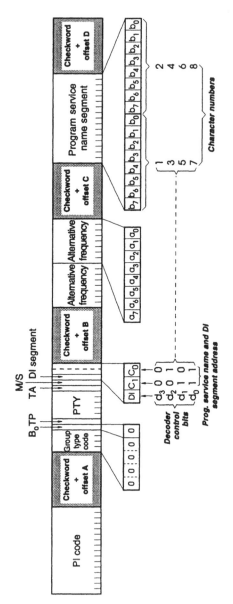

Figure 18–2 Basic tuning and switching information: Type 0A group.

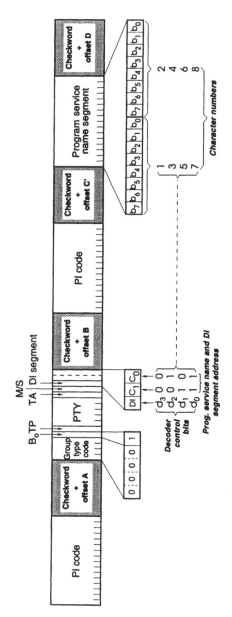

Figure 18–3   Basic tuning and switching information: Type 0B group.

- Traffic Message Channel (TMC)
- Radio Paging (RP)
- Language Identification (LI)
- Linkage codes
- Emergency Warning System (EWS)

Type 1 groups carry information about the program, as well as data about other RDS groups that need only periodic repetition, called slow-labeling codes. The slow-labeling codes also work in conjunction with other data groups to provide a particular feature, such as paging or emergency warning systems. Many of these referenced data groups can be utilized for the open data application when not flagged by the slow-labeling codes. The 1A group, shown in Figure 18.4, carries both the slow-labeling and PIN information. The 1B group, shown in Figure 18.5, carries only the PIN information. The following guidelines apply to the use of this data group:

1. When a Program Item Number is changed, a type 1 group should be repeated four times with a separation of about 0.5 seconds.
2. The unused bits in block 2 (type 1B only) are reserved for future applications.
3. Where Radio Paging is implemented in RDS, a type 1A group will be transmitted once per second, except at each full minute when it is replaced by one type 4A group.

Type 2 Groups: Radiotext

Use this group if:

- You want to transmit program-related text information, such as the current artist and song title.
- Use the A type group for messages between thirty-two and sixty-four characters in length.
- Use the B type group for messages up to thirty-two characters in length.

Type 2 groups are utilized to transmit dynamic text information related to the current program. This information may include artist and song title, station promotional material, or even emergency related information, such as a weather watch or warning. Figure 18.6 shows the format of type 2A groups and Figure 18.7 shows the format of type 2B groups. Notice that the two groups differ only in the number of text characters transmitted in each group. Due to this difference, the 2A group may carry up to sixty-four characters while the 2B group may carry up to thirty-two characters.

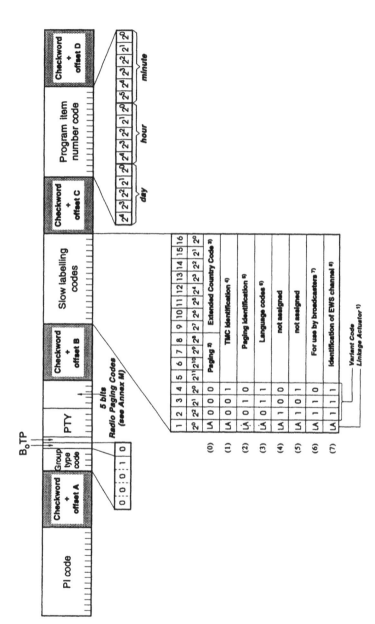

Figure 18–4 Program Item Number and slow-labeling codes: Type 1A group.

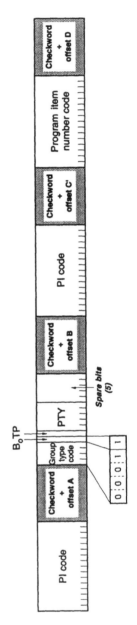

Figure 18-5 Program Number: Type 1B group.

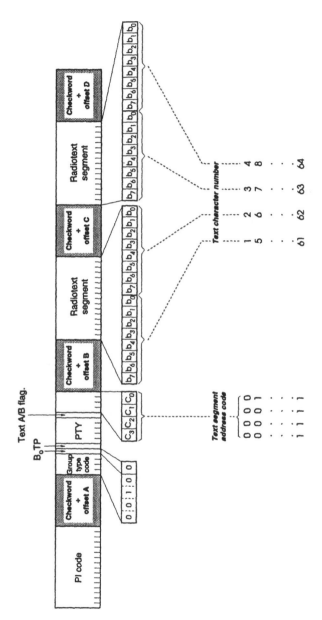

Figure 18–6 Radiotext: Type 2A group.

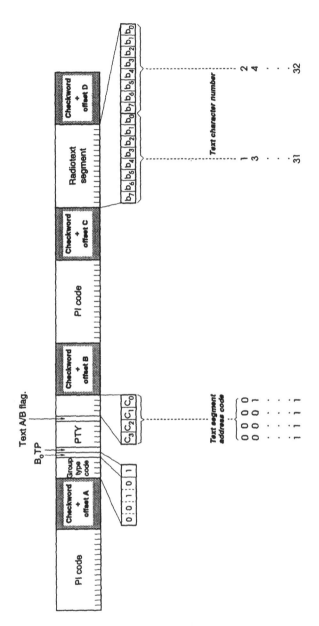

Figure 18–7 Radiotext: 2B group.

Rules of usage:

1. The four-bit text-segment address defines in the current text the position of the text segments contained in the third (version A only) and fourth blocks. Since each text segment in the version 2A groups comprises four characters, messages of up to sixty-four characters in length can be sent using this version. In version 2B groups, each text segment comprises only two characters; therefore, when using this version, the maximum message length is thirty-two characters.
2. A new text must start with segment address "0000," and there must be no gaps up to the highest-used segment address of the current message. The number of text segments is determined by the length of the message, and each message should be ended by the code 0D (Hex)—carriage return—if the current message requires less than sixteen segment addresses.
3. If a display that has fewer than sixty-four characters is used to display the radiotext message, then memory should be provided in the receiver/decoder so that elements of the message can be displayed sequentially. This may, for example, be done by displaying elements of text one at a time in sequence, or by scrolling the displayed characters of the message from right to left. Do not attempt to format the text for a particular receiver display.
4. Code 0A (Hex)—line feed—may be inserted to indicate a preferred line break.
5. It should be noted that, because of the above considerations, there is possible ambiguity between the addresses contained in version A and those contained in version B. For this reason, a mixture of type 2A and type 2B groups must not be used when transmitting any one given message.
6. An important feature of type 2 groups is the Text A/B flag contained in the second block. Two cases occur:
   a) If the receiver detects a change in the flag (from binary "0" to binary "1," or vice versa), then the whole radiotext display should be cleared and the newly received radiotext message segments should be written into the display. It is important to toggle the bit whenever the text message is changed.
   b) If the receiver detects no change in the flag, then the received text segments or characters should be written into the existing displayed message, and those segments or characters for which no update is received should be left unchanged.
7. When this application is used to transmit a thirty-two-character message, at least three type 2A groups or at least six type 2B groups should be transmitted in every two seconds.

8. Radiotext is transmitted as eight-bit characters as defined in the eight-bit code tables in Annex E of the *Standard*. The most significant bit (b_7) of each character is transmitted first.
9. The addresses of the characters increase from left to right in the display.

Type 4A Groups: Clock-Time and Date

Use this group to:

* Convey the accurate time and date to your listeners.
* Do not use the CT feature unless you accurately maintain the time within 15 seconds.
* Remember that the time offset from UTC (Universal Time Coordinated) must be changed at the beginning and end of daylight savings time.
* Provide accurate time for services, such as paging or the traffic message channel.

The transmitted clock-time and date shall be accurately set to UTC plus local offset time. Otherwise, the transmitted CT codes shall all be set to zero. Figure 18.8 shows the format of type 4A groups. When this application is used, one type 4A group will be transmitted every minute. It is not necessary to enter this group in the group structure, since most encoders automatically insert the time once per minute.

Notes on Type 4A Groups:

1. The local time is composed of Universal Time Coordinated (UTC) plus the local time offset.
2. The local time offset is expressed in multiples of half hours within the range -12 h to $+12$ h and is coded as a six-bit binary number. "0" = positive offset (East of zero degree longitude), and "1" = negative offset (West of zero degree longitude). Eastern standard time is expressed as an offset of -5 hours.
3. The clock-time group is inserted so that the minute edge will occur within ± 0.1 seconds of the end of the clock-time group.
4. The date is expressed in terms of the Modified Julian Day. Simple conversion formulas to month and day, or to week number and day of week, are given in Annex G of the *Standard*. Note that the Modified Julian Day date changes at UTC midnight, not at local midnight.
5. Accurate CT based on UTC plus local time offset must be implemented on the transmission where TMC and/or radio paging is implemented.

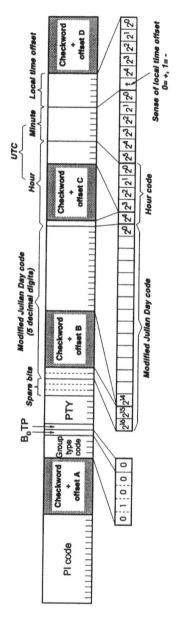

Figure 18-8 Clock-time and date transmission: Type 4A group.

Type 10A Groups: Program Type Name

Use this group to:

- Provide a better or more detailed description of the program type code that is currently selected.

Figure 18.9 shows the format of type 10A groups used for PTYN. The type 10A group allows further description of the current Program Type; for example, when using the PTY code of "Sports," a PTYN of "Football" may be indicated to give more detail about that program. PTYN must only be used to enhance Program Type information, and it must not be used for sequential information.

Notes on Type 10A Groups:

1. The A/B flag is toggled when a change is made in the PTYN being broadcast.
2. The Program Type Name (PTYN)—for display—is transmitted as eight-bit characters as defined in the eight-bit code tables in Annex E of the *Standard*. Eight characters (including spaces) are allowed for each PTYN and are transmitted as four-character segments in each type 10A group. These segments are located in the displayed PTY name by the code bit C_0 in block 2. The addresses of the characters increase from left to right in the display. The most significant bit (b_7) of each character is transmitted first.

Type 14 Groups: Enhanced Other Networks Information

Use this group to:

- Provide additional information and services about related programs, such as:
  The Program Service name, Alternate Frequencies, and Program Identification code of related stations or networks.
  Share program information such as Program Item Number, program type, and Traffic Announcements on other networks. This allows receivers to switch to affiliated stations for services such as traffic announcements or news flashes.
- Provide linkage information in order to temporarily link stations with different PI codes.
- The 14B group is to be used only to flag the beginning of a traffic announcement on a referenced network.

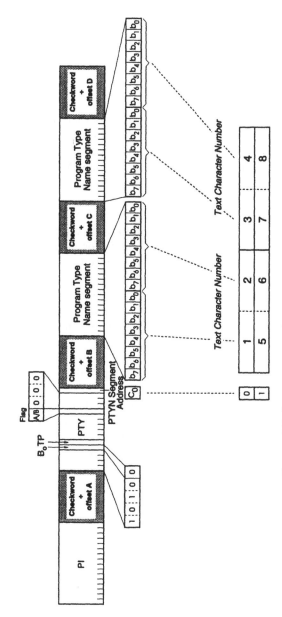

Figure 18-9 Program Type Name (PTYN): Type 10A group.

• EON will only be used by consumer receivers that are equipped with the EON feature.

Figures 18.10 and 18.11 show the format of type 14A and 14B groups. These groups are transmitted if Enhanced Other Networks (EON) information is implemented.

Rules for use:

• The A version is the normal form and shall be used for the background transmission of Enhanced Other Networks information.
• The maximum cycle time for the transmission of all data relating to all cross-referenced program services shall be less than two minutes.
• The A version has sixteen variants that may be used in any mixture and order.
• Both AF method A and B are supported. A broadcaster should choose the most appropriate AF method for each cross-referenced Program Service. Only one method can be used for each referenced program.
• The B version of a type 14 group is used to indicate a change in the status of the TA flag of a cross-referenced Program Service.
• The type 14B group is used to cause the receiver to switch to a Program Service that carries a Traffic Announcement. When a particular Program Service begins a Traffic Announcement, all transmitters that cross-reference this service via the EON feature shall broadcast as many as possible of up to eight and at least four appropriate group 14B messages within the shortest practicable period of time (at least four type 14B groups per second).
• If a transmitter cross-references to more than one traffic program with different PI(ON) via the EON feature, the time between two different references via the type 14B group has to be longer than two seconds.
• Note: Some early RDS-EON consumer receivers may need up to four correct type 14B groups for reliable functioning. Therefore, it is recommended to broadcast as many as possible of up to eight type 14B groups to ensure the detection of the switching under bad receiving conditions.

Type 15A Groups: Fast Program Service Name

Do not use this group type. This group is being phased out of the *RBDS Standard*. Broadcast encoders and receiver manufacturers should remove this group type as soon as possible. In order to maintain compatibility with existing receivers that may incorporate this feature, this group will be unavailable for use for ten years (approximately until the year 2007). The European *RDS Standard* also reserves this group for future definition, but it

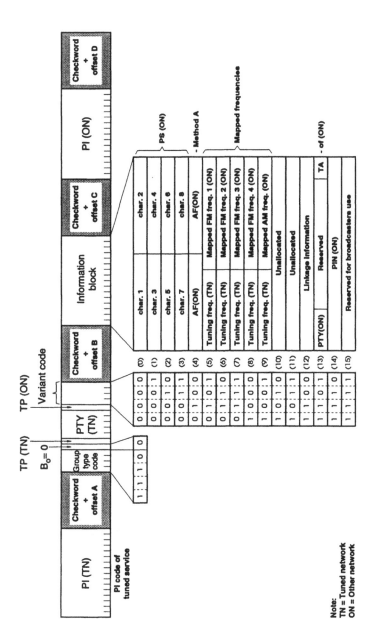

Figure 18-10 Enhanced Other Networks information: Type 14A groups.

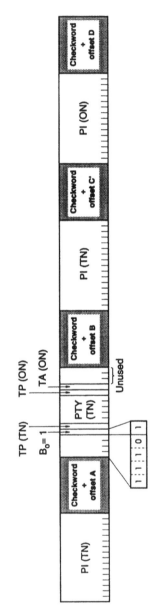

Figure 18–11 Enhanced Other Networks information: Type 14B groups.

currently makes no use of the group type. The fast PS feature allowed the transmission of four Program Service name characters per group. The Program Service name is conveyed in group types 0A and 0B at two characters per group.

Type 15B Groups: Fast Basic Tuning and Switching Information

This group is used to quickly convey to consumer receivers the PI code, traffic status (TP/TA), decoder information, and status of the Music/Speech flag. The 15B group repeats this data more frequently than type 0 groups do. 15B groups (Figure 18.12) should be used in combination with type 0 groups, but never to totally replace type 0 groups.

SERVICE-ORIENTED DATA GROUPS

The following set of ancillary data services may also be used for Open Data Applications when they are not utilized for their primary purpose. These data groups are all referenced within group type 1A slow-labeling codes.

Type 8A Groups: Traffic Message Channel or ODA

The Traffic Message Channel or TMC is an expansion of the Traffic Announcement feature. Whereas the Traffic Announcement feature is used to signal when an actual audible bulletin or update is being read, TMC conveys all relevant traffic information in a data-only format. The specification of TMC that also makes use of type 1A and/or 3A and 4A groups, is separately defined by a standard issued by the European Committee for Electrotechnical Standardization (see document prENV 12313). TMC is currently being rolled out in a large scale among several European countries. It is also being investigated for use in the U.S. as part of the Intelligent Transportation System (ITS).

Specially designed consumer receivers are required to make use of TMC information. Figure 18.13 shows the format of type 8A groups when they are used for Traffic Message Channel (TMC); if they are used for ODA, see Figure 18.22.

Type 9A Groups: Emergency Warning Systems or ODA

The use of RDS to convey ancillary emergency information is designed to be flexible so that each country may design a data format to best suit its own particular needs. For this reason, there is no predefined data that is required

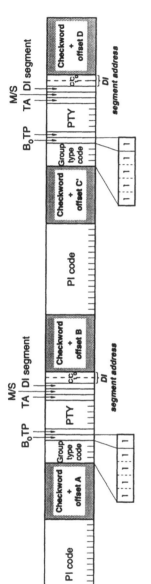

Figure 18–12   Fast basic tuning and switching information: Type 15B group.

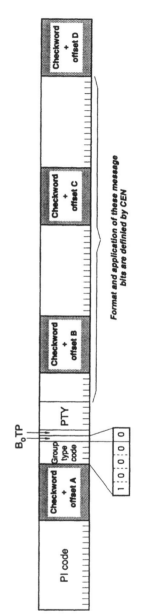

Figure 18–13   Traffic Message Channel: Type 8A group.

to be contained with group 9A. When utilized for emergency data, the EWS reference within the slow-labeling codes (group 1A) should be transmitted. Additionally, the ECC feature must be transmitted in type 1A groups when EWS is implemented. These groups are transmitted very infrequently, unless an emergency occurs or test transmissions are required. In the U.S., a special Open Data Application is available for use in both public and private emergency warning systems. SAGE Alert makes use of the 1A/9A combination within their emergency networks. Figure 18.14 shows the format of type 9A groups when used for EWS. If used for ODA, Figure 18.22 is applied.

Type 7A Group: Radio Paging or ODA

Figure 18.15 shows the format of type 7A groups when they are used for Radio Paging; if used for ODA, see Figure 18.22. The specification of RP that also makes use of type 1A, 4A, and 13A groups is given in Annex M of the *Standard*.

Type 13A Groups: Enhanced Radio Paging or ODA

The type 13A group is used to transmit the information relative to the network and the paging traffic. Its primary purpose is to provide an efficient tool for increasing the battery lifetime of the pager.

Figure 18.16 shows the format of the type 13A group. These groups are transmitted once or twice at the beginning of every interval, either after the type 4A group at the beginning of each minute, or after the first type 1A group at the beginning of each interval.

The STY code (three bits) denotes the different type 13A group subtypes; there are eight different subtypes. The specification of the relevant protocol is given in Annex M, section M.3 of the *Standard*. The type 13A group may be used for ODA when it is not used for Radio Paging; then its group structure is as shown in Figure 18.22.

DATA TRANSMISSION

Prior to the development of the Open Data Application (ODA), the transparent data and in-house data channels were the only methods for data transmission available for use. To maintain compatibility with existing services, these groups have been retained for use. When not utilized under the existing definitions, these groups may be used as ODA.

Type 5 Groups: Transparent Data Channels or ODA

Figure 18.17 shows the format of type 5A groups and Figure 18.18 the format of type 5B groups when they are used for TDC; if used for ODA, see Figures

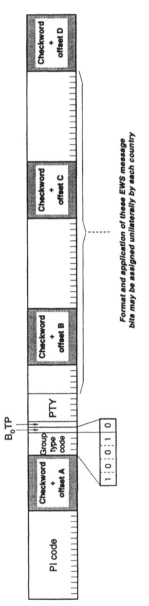

Figure 18–14 Allocation of EWS message bits: Type 9A group.

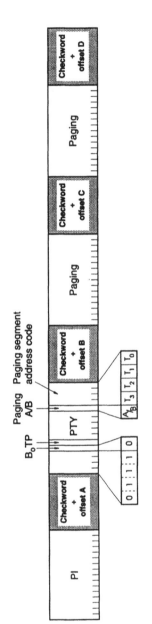

Figure 18–15 Radio Paging: Type 7A group.

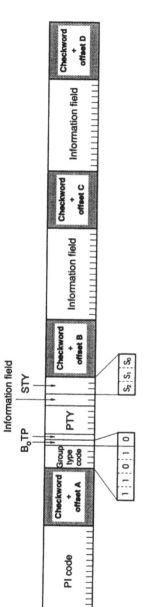

Figure 18–16 Enhanced Paging information: Type 13A group.

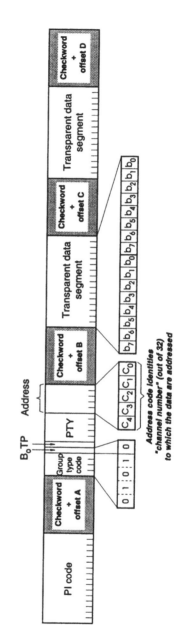

Figure 18–17 Transparent data channels: Type 5A group.

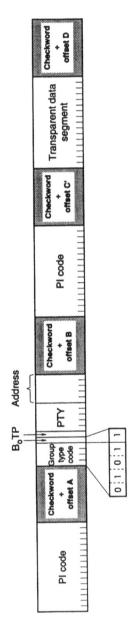

Figure 18-18 Transparent data channels: Type 5B group.

18.22 and 18.23. The five-bit address code in the second block identifies the "channel-number" (out of thirty-two) to which the data contained in blocks 3 (version A only) and 4 are addressed. Unlike the fixed-format Radiotext of type 2 groups, messages of any length and format can be sent using these channels. Display control characters (such as line-feed and carriage-return) will, of course, be sent along with the data.

These channels may be used to send alphanumeric characters or other text (including mosaic graphics), or for transmission of computer programs and similar data not for display. The repetition rate of these group types may be chosen to suit the application and the available channel capacity at the time. This group is a predecessor to the development of the Open Data Application.

Type 6 Groups: In-House (IH) Applications or ODA

Figure 18.19 shows the format of type 6A groups and the format of type 6B groups when they are used for IH. If used for ODA, see Figures 18.22 and 18.23. The contents of the unreserved bits in these groups may be defined unilaterally by the operator. Consumer receivers should ignore the in-house information coded in these groups. The repetition rate of these group types may be chosen to suit the application and the available channel capacity at the time. This group is intended for use by the broadcaster or broadcast company for its own particular uses.

Open Data Applications

The use of Open Data Applications is determined by group type 3A. This group signals the application that is being carried through the Application Identification (AID) code. AID codes are assigned by the European RDS forum and by the U.S. National Association of Broadcasters. These codes are internationally allocated and assigned for each application. Both private and public data services may be carried through the ODA. When data is contained outside group 3A, the application group type code identifies the group that carries this data. For a detailed description of the ODA, see Chapter 16.

Type 3A Groups: Application Identification for Open Data

Figure 18.20 shows the format of type 3A groups. These groups are used to identify the Open Data Application (ODA) in use on an RDS transmission.

The type 3A group conveys, to a receiver, information about which Open Data Applications are carried on a particular transmission and in which groups they will be found. The type 3A group comprises three elements: the Application Group type code used by that application, sixteen message bits for the actual ODA, and the Applications Identification (AID) code.

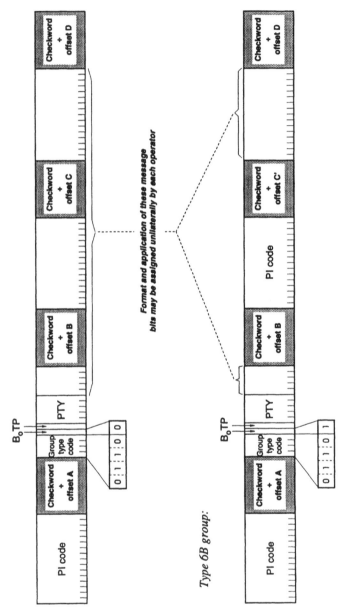

Type 6A group:

Type 6B group:

Figure 18-19 In-house applications: Type 6A and 6B groups.

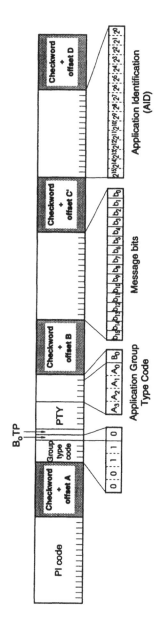

Figure 18–20 Application Identification for Open data: Type 3A group.

Group	Group type code/version				
type	A₃	A₂	A₁	A₀	B₀
3 B	0	0	1	1	1
4 B	0	1	0	0	1
7 B	0	1	1	1	1
8 B	1	0	0	0	1
9 B	1	0	0	1	1
10 B	1	0	1	0	1
11 A	1	0	1	1	0
11 B	1	0	1	1	1
12 A	1	1	0	0	0
12 B	1	1	0	0	1
13 B	1	1	0	1	1

Figure 18-21 RDS Open Data Application group types.

Open Data Application Data Groups

These groups are only to be utilized for Open Data Applications. There are two types of ODA data groups:

1. Those that can be used for ODA when not being utilized for their primary feature. These groups were identified in the previous group descriptions.
2. Those that are assigned only for ODA. Figure 18.21 identifies the groups that are only assigned for ODA.

Open Data Applications: Group Structure

Of the groups assigned only for ODA, there are two types: A version and B version groups. Open Data Applications must use the format shown in Figure 18.22 for ODA type A groups and in Figure 18.23 for ODA type B groups. Note that type B groups always carry the PI code in block 3, even

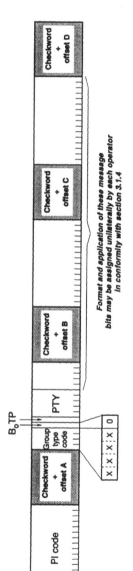

Format and application of these message bits may be assigned unilaterally by each operator in conformity with section 3.1.4

Figure 18–22 ODA type A groups.

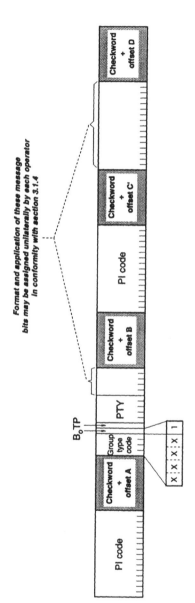

Format and application of these message bits may be assigned unilaterally by each operator in conformity with section 3.1.4

Figure 18–23 ODA type B groups.

if that group is only utilized within an ODA. This is done to ensure compatibility with consumer receivers. Receivers will always recognize the third block as the PI code for type B groups, regardless of the actual group number. To ensure backwards compatibility even an ODA must follow this structure.

RDS Broadcast Equipment

19

A wide variety of broadcast equipment exists to help you properly implement RDS at your facility. Equipment includes RDS encoding equipment, automation software, and data analyzers. Choosing the proper equipment depends on your short- and long-term needs and goals for RDS features and services. You should be familiar with the RDS feature set and have some idea of how your station will utilize RDS before choosing equipment. The basic principal is, the more you want to do, the more money it will cost. Some things just never change.

Broadcast RDS Encoders

There are at least a dozen different manufacturers of encoding equipment today. The most common form is housed in a rack mount box that is installed in the FM composite line. Most units operate with a built-in microcontroller that allows the unit to operate autonomously. Some encoders are PC-based, installing directly into a personal computer and requiring the full-time resources of the PC to operate. There are two basic types of encoders available:

1. *Consumer-based RDS encoder.* This type of encoder typically has a single input control device, such as an RS-232 port, that allows remote control of the encoder data. Generally, these units include some method of storing one or more RDS data files or records. This allows you to program the RDS data using a PC, then disconnect the

183

PC until the data needs to be changed. Many units allow multiple files to be created and stored. These files can be selected via a computer, or often through a hardware switch closure. The hardware switch closure allows remote control of the RDS data without the full-time use of a PC.

2. *Data-based RDS encoder.* A data-based encoder is designed for applications that require multiple data inputs, such as multiple RS-232 ports. Each port can be individually allocated to allow manipulation of fixed-data groups, as well as priority control. This type of encoder is almost a necessity if multiple data services are offered, such as multiple Open Data Applications. Consumer features are still supported with this type of encoder, but the design is optimized for multiple inputs to the single RDS datastream output. Making the decision of which encoder to buy will depend on the long-term goals of the station regarding RDS usage.

Static versus Dynamic RDS Data

Many RDS features are static in nature and therefore do not require automation control. Static features include: Program Service (PS) name, Program Identification (PI) code, Traffic Program (TP) bit, Decoder Information (DI) codes, and Alternate Frequency (AF) lists. Dynamic features include: Program Type (PTY) codes, Program Type Names (PTYN), Radiotext (RT), Music/Speech flag, and the Traffic Announcement (TA) bit. Almost all of the dynamic features can be implemented in a static mode, however RDS offers much more when operated dynamically.

In this book we examined the interaction between consumer receivers and the RDS data. Many features are just plain boring or of limited benefit if they are not changed to reflect the current broadcast audio program. Since the broadcast industry is market-driven, it may take some time before a full range of automation interfaces are developed. Plan now to drive your automation suppliers into the new age of interactive radio by informing them of your needs. The interface to many existing products is quite simple, so add-on costs will be comparatively small. Start small and work your way up.

Since many encoders include a hardware interface to select from the stored data records, it is possible to built an inexpensive timer-driven circuit that will select each data record for a minute or so at a time. Copy all the data records so that they are the same, then change only the Radiotext message in each record. With this ten-dollar project you have gained a Radiotext automation system that offers much more benefit than a single-fixed message. It may not be artist and song title information today, but that's next month's project!

Communicating with Encoders: The Universal Encoder Protocol (UECP)

While most encoders make use of a common hardware communication method, namely RS-232, they vary widely in the actual command-line structure. Each encoder manufacturer, therefore, has created his own interface requirements. The UECP was born out of the frustration of application vendors who are forced to supply equipment that must be able to communicate with several dozen different protocols in order to be used with all encoders. The UECP calls for a common command structure for all RDS features, but still allows support of manufacturer-specific commands.

The UECP was created after the fact, however. In other words, it was developed after most encoder manufacturers had already designed and sold equipment. Therefore, few manufacturers currently support this protocol. With all the changes that have just occurred to the *Standard* it may be possible that many more manufacturers will include the UECP with their firmware upgrades. Encoders equipped with the UECP will make life easier for you, since application vendors will only need to support one protocol. This means that all the automation vendors will have products that can communicate with your encoder, greatly expanding your available choices.

Subcarrier Frequency

The RDS subcarrier should be locked to the third harmonic of the 19 kHz pilot signal during stereo broadcasts. The frequency tolerance of the 19 kHz pilot is +/−2 Hz, making the RDS subcarrier +/−6 Hz. Monophonic transmissions must also meet the same frequency tolerance.

Phase Adjustment

In order to minimize possible interference of RDS to the main program, or vice versa, the RDS subcarrier must be phase-locked to the pilot signal (+/−10 degrees). Many encoders will automatically set the phase-lock, while less expensive models require manual adjustment. Some RDS monitors offer a phase-readout as a feature.

Injection Adjustment

The subcarrier level is specified in the range of +/−1.0 kHz to +/−7.5 kHz. This is the single most important adjustment that you must make during the encoder set-up, for it directly affects how well the RDS data is received. From a consumer point of view, the RDS data should be as reliable as the broadcast audio. It is not expected that the RDS data will be received without error where the audio is severely disturbed. Conversely, it is expected that the RDS data be received with some confidence where the audio quality is acceptable. Remember, consumer receivers will be reading the RDS data before the audio is unmuted, so reliable data is a concern. If data transmission is also a requirement of your station, then more injection is better. Many data receivers, such as pagers, have poor sensitivity performance, making high injection levels a must for reliable performance.

A Case Study of RDS Injection Levels

A study was performed in Germany that examined the effect of various injection levels on the receiver's ability to properly decode critical RDS data. The PI and PS features are essential to the operation of consumer receivers when tuning to an RDS station. The data presented here utilized no error-correction of the RDS data. The study was performed using a multipath simulator that was set to simulate a variety of listening areas. The power level of the test was set to 30 dB(pW) and performed using the average of two different mid-priced consumer automotive receivers. This power level represents the 90% coverage area for the transmitter. The study involved injection levels of 1.2 to 2.0 kHz only.

The study shows the percentage of the time that the data was properly received. As can be seen in Figure 19.1, very low injection levels offer a highly-reliable data system under normal urban and rural areas. Under adverse urban and rural areas, there is a sharp decrease in the reliability of the data. However, under these conditions, the PI code is received with better reliability than the PS name. Consumer receivers often store the PS name, displaying the stored name once the PI code is received. This helps to offset the difference in the reliability between the two when the tuned station has been previously stored. Figures 19.2 and 19.3 present the data in a graphic format. Note that an increase in RDS injection results in a near linear improvement in data transmission reliability. Therefore, it can be assumed that increases in injection over 2.0 kHz will continue to result in a linear increase in the percentage of PI and PS data received.

Monitoring RDS Broadcasts

Set-up and programming of your RDS encoder can often be made easier with the help of an RDS monitor. Several different types are commercially available for broadcaster use. The most common types are:

Dev. kHz	PI	Delta PI	PS	Delta PS
1.2	97.78	0	89.91	0
1.3	98.17	0.39	91.41	1.5
1.4	98.55	0.77	92.91	3
1.5	98.94	1.16	94.42	4.51
1.6	98.99	1.21	94.71	4.8
1.7	99.04	1.26	95	5.09
1.8	99.09	1.31	95.29	5.38
1.9	99.14	1.36	95.58	5.67
2	99.19	1.41	95.87	5.96

Rural area / Non-hilly

Dev. kHz	PI	Delta PI	PS	Delta PS
1.2	98.56	0	93.51	0
1.3	98.73	0.17	93.96	0.45
1.4	98.91	0.35	94.42	0.91
1.5	99.09	0.53	94.87	1.36
1.6	99.12	0.56	95.09	1.58
1.7	99.14	0.58	95.31	1.8
1.8	99.17	0.61	95.53	2.02
1.9	99.2	0.64	95.75	2.24
2	99.23	0.67	95.97	2.46

Typical urban area

Dev. kHz	PI	Delta PI	PS	Delta PS
1.2	70.38	0	34.69	0
1.3	72.34	1.96	36.88	2.19
1.4	74.3	3.92	39.06	4.37
1.5	76.27	5.89	41.25	6.56
1.6	77.53	7.15	43.03	8.34
1.7	78.8	8.42	44.8	10.11
1.8	80.07	9.69	46.58	11.89
1.9	81.34	10.96	48.35	13.66
2	82.61	12.23	50.13	15.44

Bad urban area

Dev. kHz	PI	Delta PI	PS	Delta PS
1.2	20.63	0	5.11	0
1.3	23.57	2.94	5.98	0.87
1.4	26.51	5.88	6.86	1.75
1.5	29.46	8.83	7.73	2.62
1.6	31.57	10.94	8.63	3.52
1.7	33.68	13.05	9.53	4.42
1.8	35.8	15.17	10.43	5.32
1.9	37.91	17.28	11.33	6.22
2	40.03	19.4	12.24	7.13

Hilly terrain

Figure 19-1 Table of results: RDS injection level versus percent received PI and PS.

Program Identfication Code Reliability

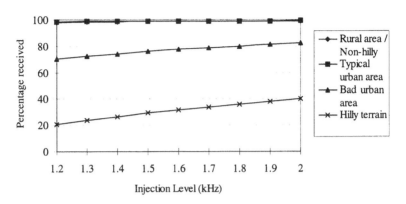

Figure 19-2 Program Identification reliability versus injection level.

Program Service Name Reliability

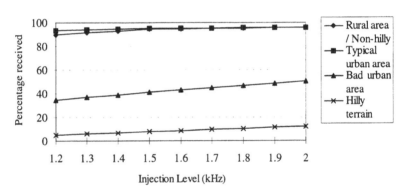

Figure 19-3 Program Service reliability versus injection level.

- *Stand-alone broadcast monitor.* This type of decoder is installed in-line after the RDS encoder or on the output of a broadcast receiver. The unit typically has a built-in display as well as a computer interface to monitor RDS data. Some units also provide RDS phase and injection level readouts for easy set-up and maintenance of the encoder's RF properties. Figure 19.4 shows a typical RDS broadcast monitor.
- *Computer software decoder.* These units are often connected directly to an RDS demodulator IC. From the RDS clock and data outputs, the software analyzes and decodes the incoming datastream. Statistical measurements, such as average and peak block error rates, are often displayed. While this type of decoder requires connection into a

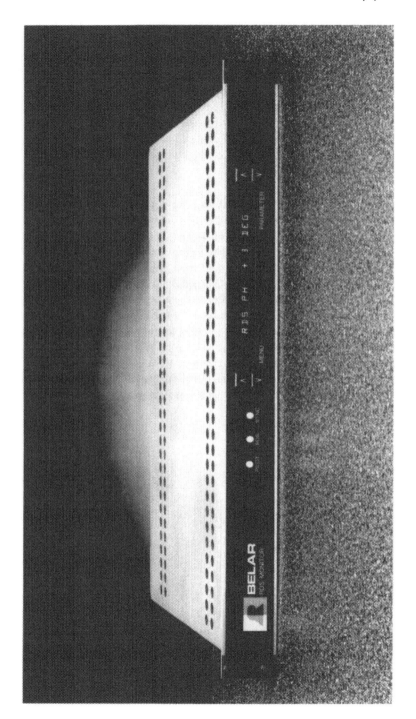

Figure 19-4 An RDS broadcast monitor. (Photo courtesy of Belar Electronics.)

receiver, it offers the advantage of portability, allowing moving vehicle monitoring of the RDS signal through an actual consumer receiver. This type of device offers real-world results by combining the RF events with the incoming data. Connection to an RDS demodulator is relatively easy. The most common manufacturers of demodulators are Philips Semiconductors and SGS Thompson. The demodulators are easily found inside almost any receiver by locating the IC nearest a 4.332 or 8.664 MHz crystal. The vendor logo can then be found and cross-referenced to a databook for that vendor, making PIN identification relatively simple.

Either device also allows you to monitor your competitor's RDS data to find out everything they are doing. By seeing the actual RDS data, programming and set-up errors are easily captured and corrected. I have also found that the software decoders are a good learning tool, especially when optimizing the data rate and throughput of all the available features. A simple method of assessing the quality of an RDS broadcast is to utilize a consumer RDS receiver. It is important to remember that some data is stored within the receiver's memory (such as the PS name). This data is instantly recalled when the receiver's preset is recalled. It is best to utilize a receiver with the Radiotext feature, since this data cannot be stored. The second-best alternative is to manually tune off and on the station's frequency and record the length of time required for the PS name to be displayed. This will yield the length of time to properly receive the PI code, since the PS name is stored in the receiver's memory.

List of Abbreviations

AF	Alternative Frequencies list
AID	Applications Identification for ODA
ARI	Autofahrer Rundfunk Information
CI	Country Identifier
CT	Clock Time and date
DI	Decoder Identification
ECC	Extended Country Code
EG	Extended Generic indicator
EON	Enhanced Other Networks information
EWS	Emergency Warning System
IH	In-House application
ILS	International Linkage Set indicator
LA	Linkage Actuator
LI	Linkage Identifier
LSN	Linkage Set Number
MBS	Mobile Search
MMBS	Modified Mobile Search
MS	Music/Speech switch
ODA	Open Data Applications
PI	Program Identification
PIN	Program Item Number
PS	Program Service name
PTY	Program Type
PTYN	Program Type Name
RBDS	Radio Broadcast Data System

RDS	Radio Data System
RP	Radio Paging
RT	Radiotext
TA	Traffic Announcement flag
TDC	Transparent Data Channels
TMC	Traffic Message Channel
TP	Traffic Program flag

Glossary of Terms

AF–Alternative Frequencies List The list(s) of Alternative Frequencies give information on the various transmitters broadcasting the same program in the same or adjacent reception area. They enable receivers equipped with a memory to store the list(s) in order to reduce the time for switching to another transmitter. This facility is particularly useful in the case of car and portable radios that will automatically stay tuned to the best quality signal available among the AF list.

CT–Clock-Time and Date Time and date codes should use Universal Time Coordinated (UTC) and Modified Julian Day (MJD), both of which are intended to update a free-running clock in a receiver. If MJD:0, the receiver should not be updated. The listener, however, will not use this information directly, and the conversion to local time and date will be made in the receiver's circuitry. CT is used as a time stamp by various RDS applications, thus it must be accurate. This feature must not be used unless it is accurately maintained.

DI–Decoder Identification and Dynamic PTY Indicator These bits indicate which possible operating modes are appropriate for use with the broadcast audio. They also indicate if PTY codes are switched dynamically.

ECC–Extended Country Code RDS uses its own country codes. The first most significant bits of the PI code carry the RDS country code. The four-bit coding structure only permits the definition of fifteen different codes, 1 to F (hex). Since there are many more countries to be identified, some countries have to share the same code, which does not permit unique identification. Hence, there is the need to use the Extended Country Code that is transmitted in Variant 0 of Block 3 in type 1A groups and that, together with the country identification in bits b_{15} to b_{12} of the PI code, render a unique combination. The ECC consists of eight bits.

EON–Enhanced Other Networks Information   This feature can be used to update the information stored in a receiver about Program Services other than the one received. Alternative Frequencies, the PS name, Traffic Program and Traffic Announcement identification, as well as Program Type and Program Item Number information can be transmitted for each of the other services. The relation to the corresponding program is established by means of the relevant Program Identification code. Linkage information, consisting of four data elements, provides the means by which several program services may be treated by the receiver as a single service during times in which a common program is carried. Linkage information also provides a mechanism to signal an extended set of related services.

EWS–Emergency Warning System   The EWS feature is intended to provide for the coding of warning messages. These messages will be broadcast only in cases of emergency and will only be evaluated by special receivers. EWS information may also be carried through a specially defined Open Data Application (see Annex Q of the *Standard*) or also through the mobile broadcast system (see Annex P of the *Standard*).

IH–In-House Application   This refers to data to be decoded only by the transmission operator. Some examples noted are identification of transmission origin, remote switching of networks, and paging of staff. The applications of coding may be decided by each operator.

MBS–Mobile Search (Mobilsokning)   This is the Swedish Telecommunications Administration (Televerket) Specification for the Swedish Public Radio Paging System. This system is operated in North America by CUE Paging Network.

MMBS–Modified Mobile Search   This is MBS modified for time multiplexing with RDS.

M/S–Music/Speech Switch   This is a two-state signal to provide information on whether music or speech is being broadcast. The signal would permit receivers to be equipped with two separate volume controls, one for music and one for speech, so that the listener could adjust the balance between them to suit his or her individual listening habits.

ODA–Open Data Applications   The Open Data Applications feature allows data applications not previously specified to be conveyed in a number of allocated groups in an RDS transmission. The groups allocated are indicated by the use of the type 3A group that is used to identify the data application to a receiver in use, in accordance with the registration details in the EBU/RDS Forum's *Open Data Applications Directory* and the *NRSC Open Data Applications Directory* (see Annex L of the *RBDS Standard*).

PI–Program Identification This information consists of a code enabling the receiver to distinguish between countries, areas in which the same program is transmitted, and the identification of the program itself. The code is not intended for direct display; it is assigned to each individual radio program to enable it to be distinguished from all other programs. One important application of this information would be to enable the receiver to search automatically for an Alternative Frequency in case of bad reception of the program to which the receiver is tuned; the criteria for the changeover to the new frequency would be the presence of a better signal having the same Program Identification code. The PI code is the equivalent of a station's unique signature.

PIN–Program Item Number The code should enable receivers and recorders designed to make use of this feature to respond to the particular Program Item(s) that the user has selected. Use is made of the scheduled program time, to which is added the day of the month in order to avoid ambiguity.

PS–Program Service Name This is the label of the Program Service, consisting of not more than eight alphanumeric characters coded in accordance with Annex E of the *Standard*, that is displayed by RDS receivers in order to inform the listener what program service is being broadcast by the station to which the receiver is tuned. An example for a name is "Radio 21." The Program Service name is not intended to be used for automatic search tuning and must not be used for giving sequential information.

PTY–Program Type This is an identification number to be transmitted with each program item and intended to specify the current Program Type within thirty-one possibilities. This code could be used for search tuning. The code will, moreover, enable suitable receivers and recorders to be preset to respond only to program items of the desired type. The last number, that is 31, is reserved for an *alarm* identification that is intended to switch on the audio signal when a receiver is operated in a waiting reception mode.

PTYN–Program Type Name The PTYN feature is used to further describe current PTY. PTYN permits the display of a more specific PTY description that the broadcaster can freely decide upon: for example, the PTY may be "4: Sport" and PTYN "Football." The PTYN is not intended to change the default eight characters of PTY, which will be used during search or wait modes, but only to show in detail the program type once tuned to a program. If the broadcaster is satisfied with a default PTY name, it is not necessary to use additional data capacity for PTYN. The Program Type Name is not intended to be used for automatic PTY selection and must not be used for giving sequential information.

RP–Radio Paging The RP feature is intended to provide radio paging using the existing VHF/FM broadcasts as a transport mechanism, thereby avoiding the need for a dedicated network of transmitters. Subscribers to a paging service will require a special pocket paging receiver in which the subscriber address code is stored. The detailed coding protocols are given in Annex M of the *Standard*.

RT–Radiotext This refers to dynamic text transmissions of up to either thirty-two or sixty-four characters in length. This text should be associated with the current audio program (e.g., artist and song title).

TA–Traffic Announcement Identification This is an on/off switching signal to indicate when a Traffic Announcement is on air. The signal could be used in receivers to:

1. switch automatically from any audio mode to the Traffic Announcement;
2. switch on the Traffic Announcement automatically when the receiver is in a waiting reception mode and the audio signal is muted; or
3. switch from one program to another one carrying a Traffic Announcement.

After the end of the Traffic Announcement the initial operating mode will be restored.

TDC–Transparent Data Channels The Transparent Data Channels consist of thirty-two channels that may be used to send any type of data.

TMC–Traffic Message Channel This feature is intended to be used for the coded transmission of traffic information. The coding is separately defined by a standard issued by CEN, prENV 12313.

TP–Traffic Program Identification This is a flag to indicate that the tuned program carries Traffic Announcements. The TP flag must only be set on programs that dynamically switch on the TA identification during Traffic Announcements. The signal shall be taken into account during automatic search tuning.

Bibliography

Beale, Terry, and Dietmar Kopitz. "RDS in Europe, RBDS in the USA: What Are the Differences and How Can Receivers Cope with Both Systems?" *European Broadcasting Union Technical Review* 255 (spring 1993).

Electronic Industries Association (EIA) and National Association of Broadcasters (NAB). *United States RBDS Standard, January 8, 1993—Specification of the Radio Broadcast Data System.* Prepared by the National Radio Systems Committee. Washington, DC: EIA/NAB, 1993.

European Broadcasting Union. *Guidelines for the Implementation of the RDS System.* Doc. Tech. 3260. Geneva, Switzerland: EBU, 1990

———. *Specification of the Radio Data System (RDS) for VHF/FM Sound Broadcasting.* Draft version 2.3. Geneva, Switzerland: EBU, 1996.

European Broadcasting Union, RDS Forum. *Comments on Program Service Name (PS)—Non-Standard Transmission of Type 0.* Geneva, Switzerland: EBU, 1996.

———. *The RDS Coverage Area in Relation to the RDS Frequency Deviation.* Prepared by the Institute for Radio Technology, Munich, Germany. Technical Report B 148/95. Geneva, Switzerland: EBU, 1996

European Committee for Electrotechnical Standardization (CENELEC). *Specifications of the Radio Data System.* EN 50067. Brussels, Belgium: CENELEC, 1992.

International Electrotechnical Commission (IEC). "Measurement of the Characteristics Relevant to Radio Data System (RDS)." Part X of *Methods of Measurement on Radio Receivers for Various Classes of Emission.* Draft IEC publication 315-X, prepared by Technical Committee No. 12: Radio Communications Sub-committee 12A: Receiving Equipment. Geneva, Switzerland: IEC, 1994.

Swedish Telecommunication Administration (Televerket). *Paging Receiver for the Radio Data System.* Doc. 1301/A694 3798 (Alternative B). Stockholm, Sweden: Televerket, 1986.

———. *Paging Receiver for the Swedish Public Radio Paging System.* Specification 76-1650-ZE. Stockholm, Sweden: Televerket, 1976.

Index